高等学校规划教材·机械工程

# 机械产品创新设计方法与实践

# The Method and Practice of Mechanical Product Innovative Design

张　涛　刘建华　赵　伟　主编

西北工業大學出版社

**图书在版编目(CIP)数据**

机械产品创新设计方法与实践/张涛，刘建华，赵伟主编. —西安：西北工业大学出版社，2013.8

ISBN 978-7-5612-3748-9

Ⅰ.①机… Ⅱ.①张… ②刘… ③赵… Ⅲ.①机械设计 Ⅳ.①TH122

中国版本图书馆 CIP 数据核字(2013)第 195140 号

**出版发行**：西北工业大学出版社
**通信地址**：西安市友谊西路 127 号　　**邮编**：710072
**电　　话**：(029)88493844　88491757
**网　　址**：www.nwpup.com
**印 刷 者**：陕西宝石兰印务有限责任公司
**开　　本**：787 mm×1 092 mm　1/16
**印　　张**：8.75
**字　　数**：208 千字
**版　　次**：2013 年 8 月第 1 版　　2013 年 8 月第 1 次印刷
**定　　价**：20.00 元

# 前　言
# Preface

产品设计作为工业设计的核心，对企业的生存和发展至关重要。为培养国际化的设计理念，了解更多的新知识、新理论，培养高水平的设计人员，促进工业设计专业的国际化，以实现我国自主设计的目标，适应经济全球化和科技革命的挑战，在工业设计专业课程中开设了“产品设计”课程的双语教学环节。

本书是根据国家教育部“大力推广使用外语讲授公共课和专业课”精神和高等学校开展双语课程改革和实践的要求，并结合多年教学实践经验编写而成的。在编写过程中，突出了以下特点：

(1)本书采用了中英文相结合方式，通过不同表述方式，释放语言限制，突破传统思维，转换思维模式，开启思路，挖掘潜能，以全新思维构架，从不同角度认识设计基本原理及思路。其中设计理论、思想及方法采用英文撰写；创新设计案例及实践则采用中文撰写，保证设计构思的清晰准确实现，加强对设计理论的认知和理解。

(2)本书内容以“设计方法—设计实例—设计创新—设计实践”为主线，阶梯递进，逐步深入，注重理论与实践紧密结合。论述语言通俗浅显、清晰简洁、条理明朗，应用实例贴近生活、易于对比和理解。

(3)本书包括国内外先进的设计理念、设计成果，可以开阔读者的视野，激发其设计与创作的欲望。

(4)本书产品设计实例范围全面，包括交通工具、家具、家电、公共设施、老年和儿童产品、日常用品、机电产品设计、健康产品等各类产品。

(5)本书的适用面广，既适用于机械类各专业，如工业设计、机械设计制造及自动化、车辆工程等专业，也适用于近机械类专业，如工业工程、电气工程及其自动化等专业。

本书的编写分工如下：刘建华负责第1章中1.1～1.4节的编写，张涛负责第1章中1.5节的编写，朱蕾负责第2章的编写，赵伟负责第3章的编写，李珂负责第4章的编写。本书由张涛、刘建华和赵伟任主编；张伟社教授负责审阅，其在审阅后提出了许多宝贵意见。在编写过程中，得到了西北工业大学出版社雷军等同志的大力支持和热忱帮助，在此一并表示感谢！

由于水平有限，书中难免会有不妥之处，恳请读者批评指正。

编　者

2013年3月

# 目 录
# Contents

# 第1章 通用设计理论
# Universal Design Principle

## 1.1 可视性设计准则 Perceived principles

### 1.1.1 外观式设计准则 The aspect

1．Color combinations

Color defines the world and gives different from the objects. Color is the one of the most useful and powerful design tools.

(1)Physics of color

Where does color come from? Color derives from the spectrum of light interacting in the eye with the spectral sensitivities of the light receptors. Color can only exist when three components are present：a viewer，an object and light. Pure white light contains all colors in the visible spectrum，but it is perceived as colorless. When white light hits an object，it reflects some colors which contribute to the viewer's perception of color.

And color has three components，they are hue，saturation，brightness to show differences between things in the world. For instance，hue can tell the ripe tomato or not，and saturation can tell the milk coffee or not，and brightness can tell the difference between the sun and the moon (see Fig. 1－1 ).

Fig. 1－1 The examples about three components

1) Hue

In general，"color" often is talked about hue，which is a component of color to be talked about most. It indicates that a color looks red，green，blue，yellow，orange，etc，appeared on the color wheel. As an example，look at the line of different hues in Fig. 1－2. It looks like colors from a rainbow.

2)Saturation

Saturation(see Fig. 1 - 3) is the amount of grey in a hue and represents how pure a color is. As an example of saturation, think about what happens to add milk in coffee. Just add a little milk and stir it up. It's deep brown. Then add more milk, it's still brown, but not as deep. Add more and more milk, the mix would get more faded until it looks almost white.

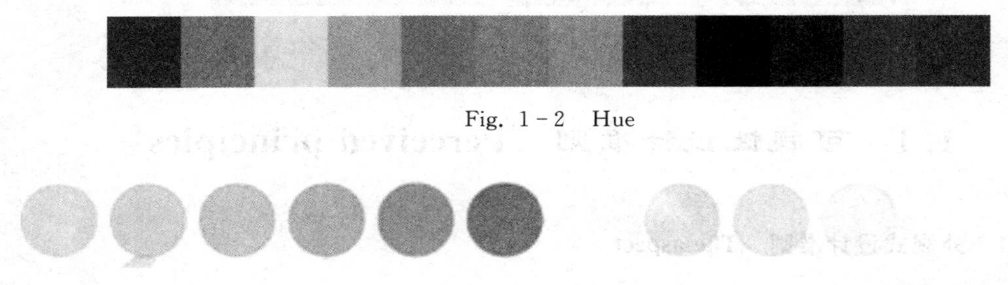

Fig. 1 - 2 Hue

Fig. 1 - 3 Saturation

The saturation increases, the amount of gray decreases. The color wheel in the middle is the purest color. A color with more gray is considered less saturated, while a bright color with very little gray, is considered highly saturated.

The saturation of a color can affect our emotion. More saturated colors are considered more bold and tied to emotions, and are perceived as more exciting and dynamic. In advertisements, saturated colors could be often used in order to catch the attention of readers and viewers. Unsaturated colors are considered softer and less striking, and are perceived as more restful and peaceful.

Then using saturated colors could attract attention, and using desaturated colors should be more performance and efficiency. Generally, desaturated dark colors are perceived as serious; and desaturated bright colors are perceived as friendly.

3)Brightness

Brightness (see Fig. 1 - 4) is the amount of white in a hue and shows how strong a color is. For example, the sun has a high brightness, while a birthday candle has a low brightness. Colors like whites and yellows have a high brightness. Colors like browns and grays have a medium brightness. Colors like black have a low brightness.

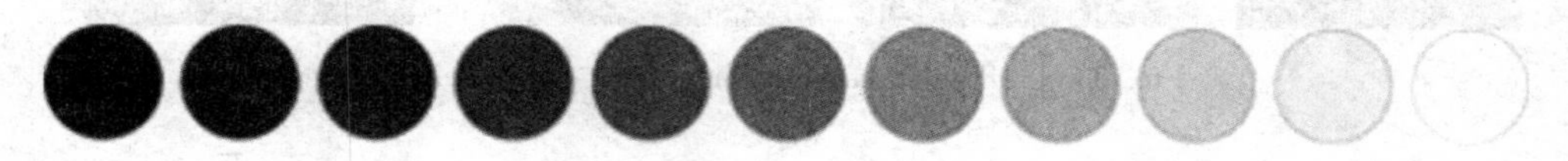

Fig. 1 - 4 Brightness

Among three components, hue is most important to seeing color. People should be more sensitive to changes in hue than to changes in brightness, and more sensitive to changes in brightness than to changes in saturation. Traffic lights use the different hues of green, yellow, and red instead of a dark red, a medium red, and a bright red.

4)Meaning of colors

Colors are non-verbal communication. Over 80% of visual information is related to color. In design, it is helpful to know how a person perceives certain colors and the color meanings.

Symbolism of color is varying from culture to culture. And different cultures attach different meanings to colors. For example, in European culture white is believed to signify marriage, angels and peace, but in the Orient white is the traditional color of mourning and death.

(2) Guidelines about Color combinations

There are the four ways to combine with colors.

1) Analogous — adjacent color

Analogous color combinations (see Fig. 1-5(a)) use colors that are next to each other on the color wheel.

2)Complement — opposing color

Complementary color combinations (see Fig. 1-5(b)) use two colors that are directly across from each other on the color wheel. These opposing colors create maximum contrast and maximum stability.

3)Quadratic

Quadratic color combinations (see Fig. 1-5(c)) use colors at the corners of a square or rectangle circumscribed in the color wheel.

4)Triadic

Triadic color combinations (see Fig. 1-5(d)) use colors at the corners of equilateral triangle circumscribed in the color wheel.

Fig. 1-5　Color combinations

(3) Epilogue

Color can attract attention, convey feeling and enhance aesthetics. And colors can also cause fatigue, increase stress, damage eyesight, increase possible worker errors. So if applied improperly, colors can seriously harm the intention of a design.

2. Rules of visibility

(1) Definition

The efficiency of a system should be improved if the users are allowed to see the information and the performing status. In general, a green light indicates that the status of the device is very well, and a red light indicates that there is failure. For example, red light shows that the printer is out of paper (see Fig. 1-6(a)); illuminated controls could be used to indicate controls that are currently available (see Fig. 1-6(b)).

In simple systems, the rules of visibility is easily realized that all controls and status are visible, but it should be perhaps difficult in complex system.

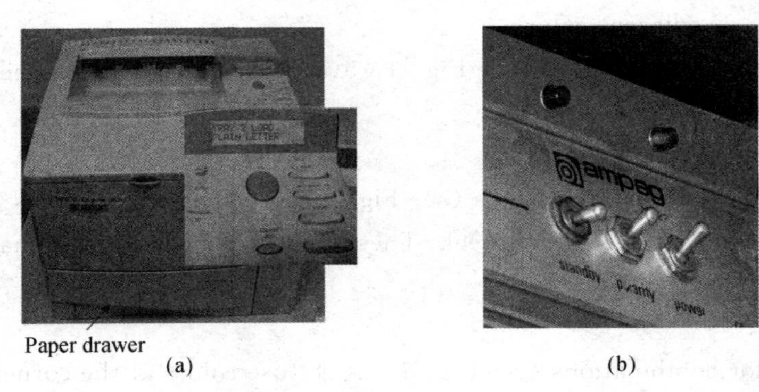

(a) (b)

Fig. 1-6 Rule of visibility

(2) The solutions

In a complex system, there are three steps for the maximize visibility. At first, the designer must consider the number of controls and information, then puts controls and information into logical categories, in final, hides some controls within a parent control, where these controls remain concealed until the parent control is activated.

3. Affordance

(1) Deduction

Afford is verb, be able to spare or give up something or be the cause or source of something. Then what is affordance?

And in 1979, Mr. James Jerome Gibson, who was an American perceptual psychologist, brought about Affordance theory. Affordance theory states that the world is perceived in terms of object shapes, spatial relationships and possibilities for action.

Mr. Gibson said, a possibility for action afforded to a perceiver by an object. And the affordances of an object depend upon the perceiver and the characteristics of the object. For example, a small stream affords such actions as crossing to an adult, but to an infant it affords jumping.

Then in 1988, Mr. Donald Norman introduced the term affordance to design in his book *Psychology of Everyday Things*. He said, "An aspect of an object which makes it obvious how it should be used" and "Keep the perceptions of the user in the mind." And that is to let object talks.

(2) Definition

At first, what is that in Fig. 1-7(a)? It is maybe a piece of glass or window? It is not sure and it is lack of affordance. But in Fig. 1-7(b), it is a blackboard, and this sure makes it obvious! And this is design with affordance.

(a) (b)

Fig. 1-7 What is that?

Affordance is a quality in which the physical characteristics of an object or environment afford its function. Physical characteristics are primarily physical in nature, such as area, hardness, smoothness, weight, shape, volume, color, etc.

Firstly, some physical characteristics are better suited for some functions than others. For example, circle is better suited than square for rolling, and stairs are better suited than fences for climbing.

Secondly, the design with affordance corresponding to its function will perform more efficiently and will be good usability. For example, the door with a flat plate affords pushing, and the door with a handle affords both pulling in Fig. 1-8(a). If open the door with a plate only by pushing and open the door with a handle only by pulling, the affordance of the designs corresponds to the way in which the door can be conveniently used. Then the sign of "pull" in Fig. 1-8(b) is superfluous. But if door with handle is designed to open only by pushing in Fig. 1-8(c), the affordance of the design conflicts with the door's function, and it would be not convenient.

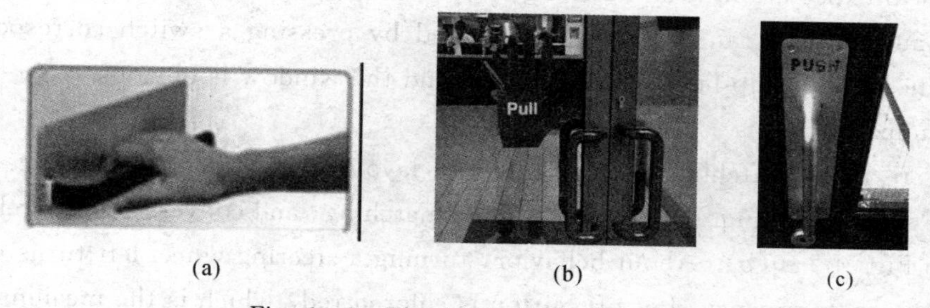


(a) (b) (c)

Fig. 1-8 The door with the handle

Thirdly, using images of common physical objects and environments can improve the usability of a design. There are many designs of images in computer software. For example, the "desktop" used by computer operating systems has the function of office desk. And on a computer screen, three dimensional buttons with the physical characteristics of buttons can afford pressing. In addition, other items, such as the file folder, rubbish can, and so on, are same with ones in the real world, and their functions are also the same with those in the real world.

(3) Practical application

Chinese abacus (see Fig. 1-9(a)), which is a calculator that performs arithmetic functions by manually sliding counters on rods or in grooves, is a typical ancient design with good affordance. And the another design is the old typewriter (see Fig. 1-9(b)) is a mechanical or electromechanical device with a set of "keys" that, when pressed, cause characters to be printed on a medium, usually paper, and it is also a typical ancient design with good affordance. And with opposing male and female surfaces and featureless sides, Legos naturally afford plugging into one another (see Fig. 1-9(c)). Besides, there are other designs such as cellphone, faucet, and so on.

(a) (b) (c)

Fig. 1-9 Some designs of good affordance

4. Matching

(1) Definition

Matching is a relationship between controls and their functions. Good matching should enhance usability.

Press a button, and twist a knob, some kind of effects should be happen. The effect corresponds to expectation, the matching is considered to be good. For example, a power window of automobiles can be raised or lowered by pressing a switch corresponds to the window. The relationship between the control and the window is obvious.

(2) The pattern

There are three matching modes, they are layout, behavior, or meaning. About the layout, the stovetop in Fig. 1-10(a) is a good matching, and conversely, the relationship is not clear in Fig. 1-10(b). About behavior, turning a steering wheel left turns the car left; and meaning, an emergency shut-off button is colored red, which is the meaning of stop.

(3) Epilogue

Be sure positions and behaviors of controls correspond to the layout and behavior of the system. Simple matching relationship is easy to use. Avoid using a single control for multiple functions.

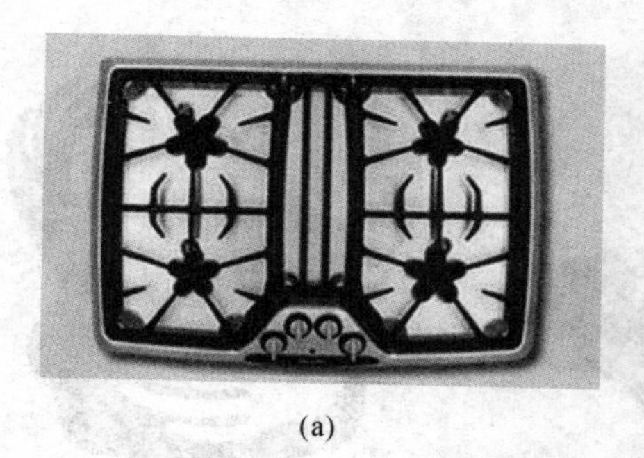

(a)

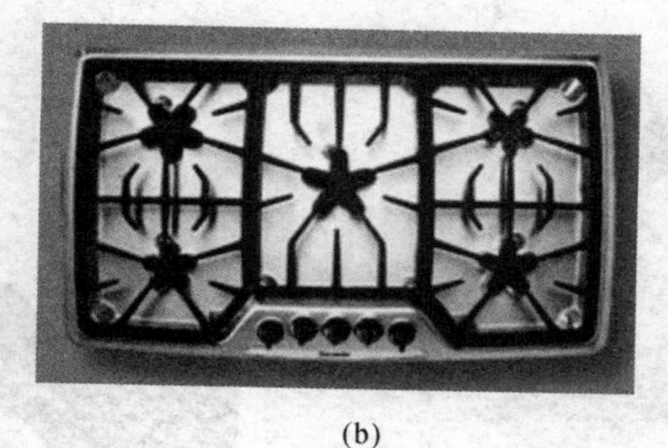

(b)

Fig. 1－10　Layout

(a) good; (b) poor

5. Figure-Ground relationship

(1) Definition

The figure-ground relationship is one of Gestalt principles of perception. Every perceived thing contains figure and ground. Figure is objects of focus. And ground is the rest of the field. Elements are perceived as either figures or ground. For example, is it a white circle on a black square or a black square with a round hole in Fig. 1－11?

There was an old joke in Soviet Russia about a guard at the factory gate who at the end of every day saw a worker walking out with a wheelbarrow full of straw. Every day he thoroughly searched the contents of the wheelbarrow, but he never found anything but straw. One day he asked the worker, "What do you gain by taking home all that straw?" "The wheelbarrows."

Fig. 1－11　Figure and ground

This paper is about the straw and the wheelbarrow, about shifting attention from figure to ground or, rather, about turning into figure what is usually perceived as ground. We are used to think of the load as "figure"; the wheelbarrow is only "ground", merely an instrument. Our default interest is in the act, not in the instrument.

(2) Figure and Ground

The part of a composition paid attention to is called figure. The figure is also called a positive shape. In different composition there may be one or several things to be figure. Everything that is not figure is ground.

This relationship can be demonstrated with both visual stimuli, such as photographs, and auditory stimuli, such as soundtracks with dialog and background music.

(3) The relationship

If the figure and ground of an object are clear, the relationship is stable; the figure element attracts more attention. In unstable figure-ground relationship, the relationship is ambiguous. At times, they could become figure from one to another thing in turn (see Fig. 1－12).

Fig. 1 - 12 The unstable figure-ground relationship

In general, the figure has a definite shape, whereas the ground is shapeless. And the figure is perceived to be closer, but the ground is perceived to be farther away in space. Things below a horizon line and the lower regions are more likely to be perceived as figures, whereas elements above a horizon line and the upper regions are more likely to be perceived as ground (see Fig. 1 - 13).

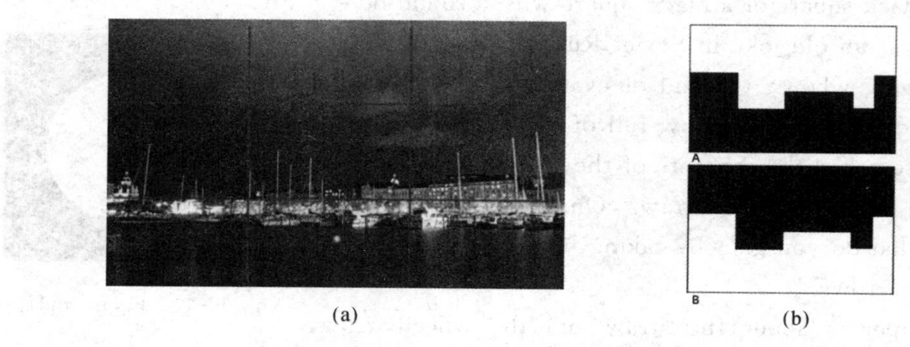


(a) (b)

Fig. 1 - 13 The relationship in the photograph

In addition to, figure-ground reversals (see Fig. 1 - 14) create a delightful "surprise" in the viewer's eye. People are unconsciously influenced by the ground even when they are consciously only aware of the figure.

Fig. 1 - 14 Figure-ground reversals

6. Arrangement

(1) Definition

Placement of elements in orderly way is aesthetic and united, such as circular, square, row, column, or spiral. And the rows and columns of people or things can clarify their relationship, for example, a formation of aircrafts in flight, or troops.

(2) The pattern

In paper or text, left-aligned, right-aligned or center-aligned text blocks are often used (see Fig. 1-15(a)), and justified text is more used than unjustified text. Sometimes to detect different from other text blocks, inclining certain angles of text block could be used (see Fig. 1-15(b)).

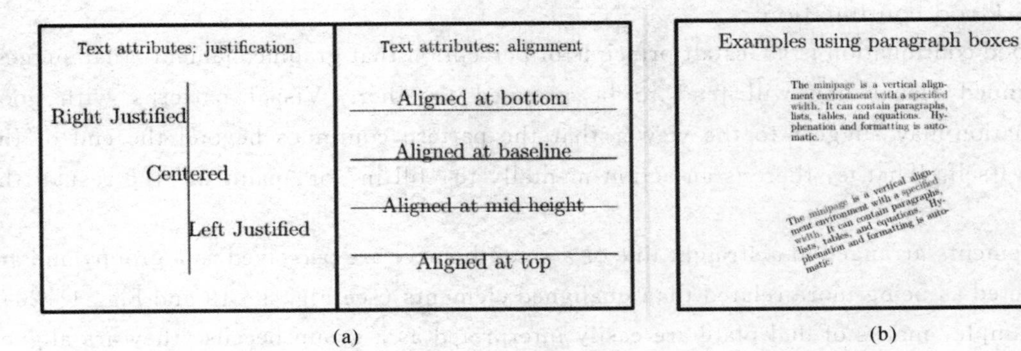


(a) (b)

Fig. 1-15 Aligned text

### 1.1.2 内涵式设计准则 The connotation

1. Consistency

Similar systems are exhibited by similar ways. Then the systems should be easy to know and use. For example, there are the same typefaces, color schemes, menus, staff uniforms, architecture in every Kentucky Fried Chicken (see Fig. 1-16).

Fig. 1-16 Kentucky Fried Chicken

There are two modes of consistency, imagery consistency and applied consistency. Imagery consistency refers to consistency of style and appearance, such as a company logo (see Fig. 1-17 and 1-18), product brand, and feature of building, etc. Applied consistency refers to consistency of meaning and action, such as traffic light, buttons of video player, and so on.

Using consistency can simplify usability and ease of learning.

Fig. 1-17 Kodak logo

Fig. 1-18 Color Gardens logo

2. Good continuation

Good continuation is a Gestalt principle of perception that graphic elements that suggest a continued visual line will tend to be grouped together. Visual patterns with good continuation may suggest to the viewer that the pattern continues beyond the end of the pattern itself. That is, there is an action mentally to "fill in" or "paint in" the rest of the pattern.

Elements arranged in a straight line or a smooth curve are perceived as a group, and are interpreted as being more related than unaligned elements (see Fig. 1-19 and Fig. 1-20). For example, marks of dial plate are easily interpreted as a group because they are aligned along a circular path. Good continuation is also important in the design of tables, especially in the alignment of columns. Readers should not look down a column to see the good continuation broken by a rule line that is intended to frame a subheading.

Using good continuation in a design, it asserts that relatedness between elements is clear and not interrupted.

Fig. 1-19 Corner of square

Fig. 1-20 Fold line

3. Five Hat Racks

The term hat racks is built on an analogy, and hats are information and racks are the ways to organize information. In 1976, Mr. Richard Saul Wurman defines the term "information architect", and the concept of the "Five Hat Racks" was originally developed in his book "Information Anxiety", indicating that the information explosion really is dramatic. However, in modern, this is not to mention the extremely dramatic expansion of electronic information on the Internet, which is probably doubling the production of information every four years. Information may be infinite, however the organization of information is finite. "Five Hat Racks" principle can effectively help to organize information.

Five Hat Racks are ways of Location, Alphabet, Time, Category, or Hierarchy. The five hat racks principle asserts that there are a limited number of organizational strategies.

Location refers to organization by geographical or spatial reference. Examples include emergency exit maps and travel guides. Alphabetical refers to organization by alphabetical sequence. Examples include dictionaries and encyclopedias. Time refers to organization by chronological sequence. Examples include historical timelines and TV Guide schedules. Category refers to organization by similarity or relatedness. Examples include areas of study in a college catalog, and types of retail merchandise on a Web site. Hierarchy refers to organization by magnitude. Examples include baseball batting averages and Internet search engine results.

## 1.2 认知性设计准则 Cognized principles

### 1.2.1 心智型设计准则 The mental

1. Mental model

(1) Definition

The term "mental model" was first defined by Craik in his book "The Nature of Explanation". In his book, Mr. Craik said that the mind constructs a "small-scale models" of reality that it uses to reason, to anticipate events and to underlie explanation. People understand and interact with environments based on mental representations developed from experience. But the concept couldn't be paid attention to until cognitive science appear. At present, the mental model has been used in many contexts and for many purposes.

Then in design, the use of mental models was popularized in the Human-Computer Interaction design field. Donald Norman in his book *Psychology of Everyday Things* used mental model to describe respectively the designer's mental model and user's mental model. A device is designed on the basis of the designer's mental model, the user forms a mental model through interaction of the device. In other words, the designer materializes his mental model of a given design, which becomes the only means of conveying his mental model to the user. Thus, users do not only interpret the visible parts of the device but also guess what goes on happening (see Fig. 1-21). Metal model is mental expression of devices, systems and environments derived from experience. User understands systems and environments, and interacts with them, through comparing the outcomes of his mental model with real-world systems and environments. If the imagine that consciously or unconsciously form from our experiences correspond to with real outcomes, the mental model is accurate and complete, on the contrary, the mental model is inaccurate or incomplete.

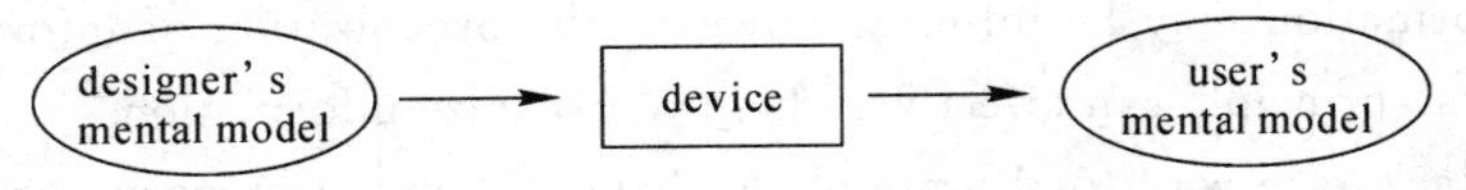


Fig. 1-21 Designer's and user's mental model

(2) The different

When a user participates in an event, the user should evaluate his expectations of the event with their realization of the event. Then before the user will participate in an event, there are three questions guide their participation. What does the user want? How does he try to achieve it? What's his action? What does he expect to happen?

At first, the user has an idea about something they want or need. The next, he has an idea about his course of action. Then they have an expectation that his action will let them realize his want. Therefore what the user wants is the goal. Since this is an idea they have, this as a "peak". And he select an approach to be close to the thing he wants, this is "path". Finally, he has an expectation of what will happen, this is "sunshine" (see Fig. 1-22). An expectation can't be formulated without a path, and the path can't be formulated without the initial goal. The goal, approach, and expectation constitute the user's mental model.

However, the designer will think more of how the device will perform in this event.

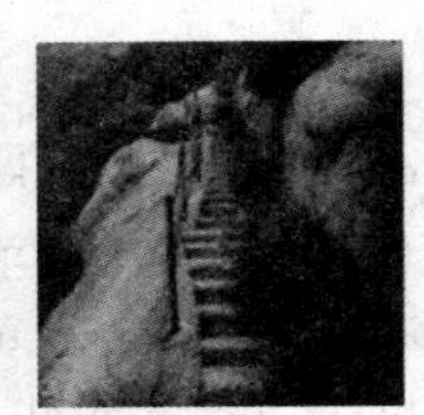
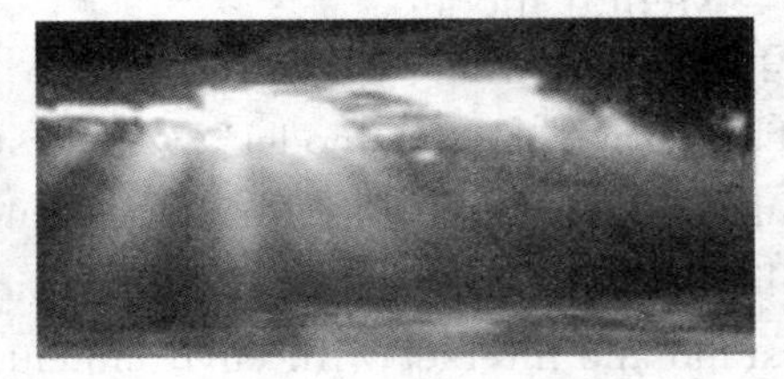

Fig. 1-22 Peak, path and sunshine

(3) The expatiation

With regards to design, there are two modes of mental models: design models and use models. Design models is the mental models of how systems work, use models is the mental models of how users interact with systems.

Designers know much about how a system works, and possess very complete and detailed design models. Conversely, users should have more complete and accurate use models than designers through use and accumulated experience. Therefore for enhancing usability of system, designers should obtain accurate and complete use models through personal use of the system, usability testing, or observation of user interacting with the system. The designer using the system or testing will learn the problems of interaction that appear when the system is used by people who are unfamiliar with it. Then the most effective method is to watch people use the design and take note of how they use it. Observing users, designers can acquire accurate information about how people interact with systems.

## 2. Operant Conditioning

(1) Deduction

Thorndike had researched on animal learning at early the twentieth century and had an enormous influence on the experimental psychology after that time. John Watson promoted the behavioristic approach in the 1920s. Probably B. F. Skinner was the well-known experimental psychologist at the twentieth century. Mr. Skinner continued Thorndike's research on instrumental learning and named it as "operant conditioning" in the 1930s. Skinner thought that individuals learn new behaviors that "operate on" the environment and behaviors that cause the individuals to experience environmental stimuli. Skinner started his experiment on the behavior of animals and observed the relationship between stimuli and response.

Skinner performed his experiments by "Skinner boxes" in Fig. 1-23. The Skinner box can realize that would automatically dispense food pellets and electric shocks. A rat or pigeon is put in the box and learns to press a lever or push a button in order to receive stimuli such as food or water. The lever press or button-push leads to the consequence, however, only when preceded by a light, voice, or other stimulus. Skinner wanted that the results of research on his Skinner boxes could apply to human behavior.

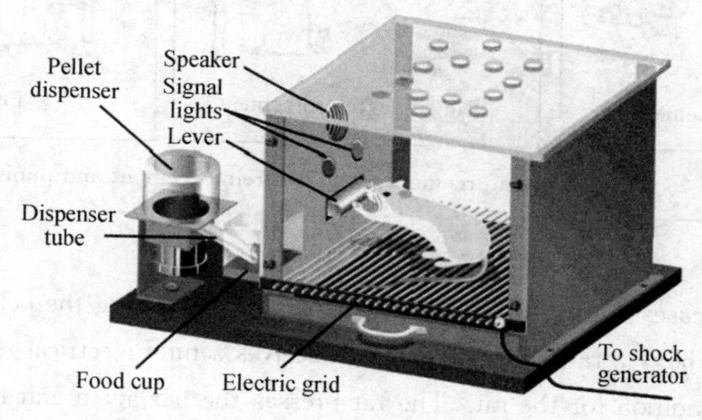


Fig. 1-23　A rat being operant conditioned to press a lever in a Skinner box

(2) The expatiation

Operant conditioning is a technique of behavior adjustments by rewards and punishment, so as to reinforce desired behaviors, and ignore or punish undesired behaviors. Operant conditioning can apply to animal training, instructional design, game or gambling devices. There are two basic operant conditioning, reinforcement and punishment with a positive or negative condition (see Fig. 1-24).

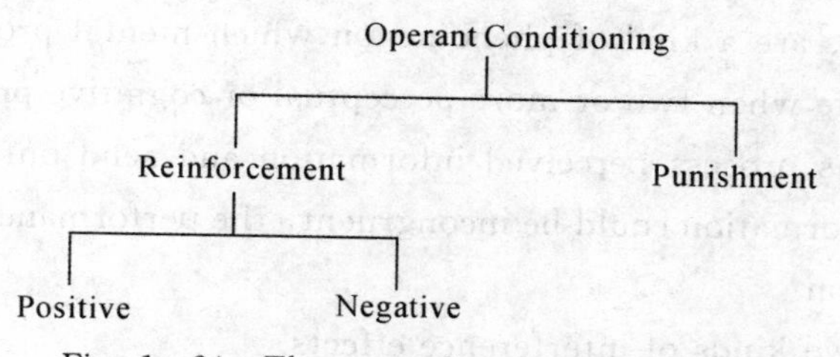


Fig. 1-24　The types of operant conditioning

1) Reinforcement — positive reinforcement and negative reinforcement

It is the process in which a behavior is strengthened, and more likely to happen again.

Positive reinforcement increases the probability of a behavior by associating the behavior with a positive condition. For example, to press the lever results in positive visual and auditory feedback, and a possible reward. In Skinner box, a hungry rat presses a bar in its cage and receives food. The food is a positive condition for the hungry rat. The rat presses the bar again, and again receives food. The rat's behavior of pressing the bar is strengthened by the consequence of receiving food.

Negative reinforcement increases the probability of a behavior by associating the behavior with the stopping or avoiding of a negative condition. In Skinner box, a rat is placed and immediately receives a mild electrical shock on its feet. The shock is a negative condition for the rat. The rat presses a bar and the shock stops. The rat receives another shock, presses the bar again, and again the shock stops. The rat's behavior of pressing the bar is strengthened by the consequence of stopping the shock.

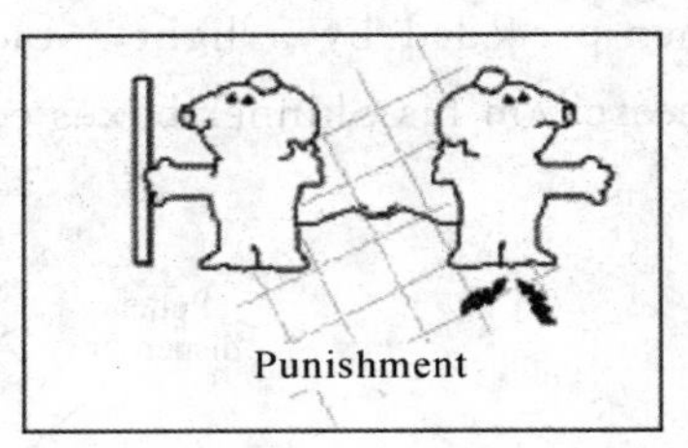


Fig. 1－25 Positive reinforcement, negative reinforcement and punishment

2) Punishment

Punishment decreases the probability of a behavior by associating the behavior with a negative condition. In Skinner box, a rat presses a bar and receives a mild electrical shock on its feet. The shock is a negative condition for the rat. The rat presses the bar again and receives a shock. The rat's behavior of pressing the bar is weakened by the consequence of receiving a shock.

3) Epilogue

Use operant conditioning in design when behavioral change is required. In general, Positive and negative reinforcement should be used instead of punishment whenever possible. Punishment should remove a behavior, so it should not be used at all.

3. Interference effects

(1) Definition

Interference effects are a kind of phenomenon which mental processing should become slower and less accurate when two or more perceptual or cognitive processes are in conflict. Human mental systems process perceived information and send out execution instruction. But if the perceived information could be incongruent, the performance should become weak.

(2) The expatiation

There are often two kinds of interference effects.

1) Stroop Interference. An irrelevant aspect of a stimulus triggers a mental process that interferes, with processes involving a relevant aspect of the stimulus. Stroop interference refers to the fact that the time it takes to name the color of words is greater when the meaning and color of the words conflict in Fig. 1-26.

The effect had previously been published in 1929 in German, and is named after John Ridley Stroop who first published the effect in English in 1935. The original paper has been one of the most cited papers in the history of experimental psychology.

For example,in the Fig. 1-27, reading quickly from top to bottom, left to right, say out loud the color of the ink of each word. If the first inclination was to read the words, "blue, pink, grey ..." rather than the colors they're printed in, "red, red, green, blue...", it is just experienced interference.

橙 蓝 绿 红
橙 蓝 绿 红

Fig. 1-26　The comparison

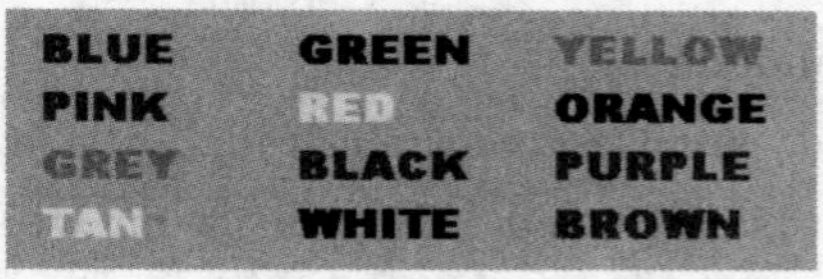

BLUE　GREEN　YELLOW
PINK　RED　ORANGE
GREY　BLACK　PURPLE
TAN　WHITE　BROWN

Fig. 1-27　Words and colors

When one person looks at one word, if he knows both its color and meaning, these two pieces of evidence should be in conflict, he has to make a choice. In general, people think that a word meaning is more important than ink color, so interference occurs when he tries to pay attention only to the ink color. The time it takes to name the color of words is greater when the meaning and color of the words conflict.

2) Garner Interference. An irrelevant variation of a stimulus triggers a mental process that interferes with processes involving a relevant aspect of the stimulus. In Fig. 1-28, naming the column of shapes that stands alone in trial 1 is easier than naming either of the columns located together in trial 2. The close proximity of the columns results in the activation of mental processes for naming proximal shapes, creating interference. The time it takes to name shapes is greater when they are presented next to shapes that change with each presentation.

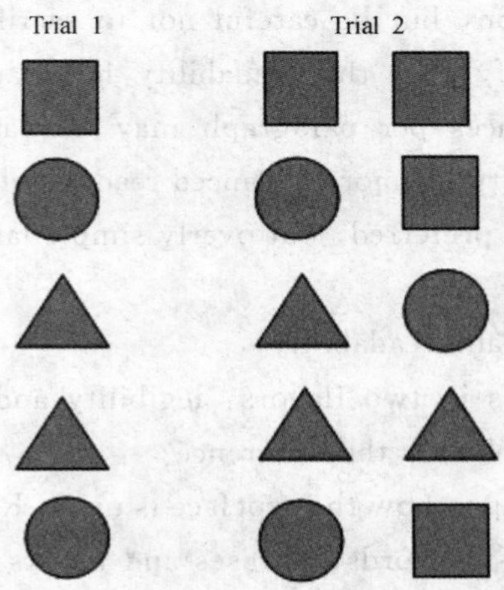


Fig. 1-28　Words and shapes

In design, the efficiency of performance should be enhanced by minimizing interference effects.

### 1.2.2 知识型设计准则 Learned from design

1. Readability

(1) Definition

Readability is the ease with which a passage can be read and understood, and is also the degree to which paper can be understood. Readability is determined by factors such as word length, number of special words, sentence length, and number of characters in a sentence. And complex information requires the simplest presentation possible, so that the focus is on the information rather than the way it is presented. Then readability is very important in design.

(2) Guidelines

It is necessary to consider readability in designs of writing and to use the simplest way possible. It's right and classy black on white is way behind according to readability study performed by Alyson L. Hill from Department of Psychology of Stephen F. Austin. At paper's webpage, text is easiest to read when the font text color and the background color are in high contrast. One of the most favorable the colors combination is which green on yellow is the best color scheme in the most conditions. And Times New Roman on average is much better than Arial (see Fig. 1-29).

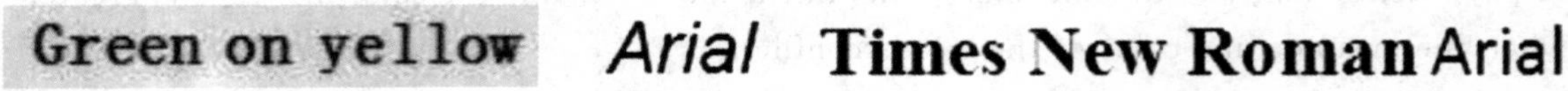

Green on yellow *Arial* Times New Roman Arial

Fig. 1-29 Colors and font of words

There are some guidelines for enhancing readability.

Generally, use active voice, but consider passive voice if the emphasis is on the message and not the messenger. Keep sentence length appropriate for the intended reader. Omit needless words and punctuation, but be careful not to sacrifice meaning or clarity in the process. At same time, verify that the readability level approximates the level of the intended reader. More sentences per paragraph may increase readability for lower-level readers, but frustrate readability for more advanced readers who are distracted by the lack of continuity. Simple language is preferred, but overly simple language obscures meaning.

2. Legibility

(1) Different of legibility and readability

Typographic clarity comes in two flavors: legibility and readability. Readability and legibility are often confused. What's the difference?

Readability is dependent upon how the typeface is used. Readability is about typography and can be a gauge of how easily words, phrases and blocks of copy can be read. On the other hand, Legibility is a function of typeface design. It's an informal measure of how easy it is to distinguish one letter from another in a particular typeface. Legibility is the degree at which glyphs and vocabulary are understandable or readable based on appearance. And legibility describes how easily or comfortably a typeset text can be read.

(2) Guidelines

In the visual clarity of text, it generally is used to the size, typeface, and spacing of the characters. The following guidelines address common issues regarding text legibility.

1) Size. For printed text, standard 9 to 12 point type is considered optimal.

2) Typeface. In Chinese, Song typeface is often used in formal document, and regular script is selected based on aesthetic preference sometimes. In western language, Times New Roman is often used.

3) Spacing. For 9 to 12 point type, set leading to the type size plus 1 to 4 points. Proportionally spaced typefaces are preferred over monospaced.

### 3. Advance Organizer

Advance organizers were popularized by D. P. Ausubel in 1960s. According to Ausubel's theory, the teacher should be responsible for organizing and presenting what is to be learned. Advance organizers are cognitive instructional strategy that helps people understand new information in terms of what they already know.

Advance organizers present information, prior to new material to help facilitate learning and understanding, which is advance organization that can be provided by words spoken or written, diagrams and charts, photographs, or models of products. They are distinct from overviews and summaries.

There are two main types of advance organizers: expository and comparative.

(1) Expository advance organizers

If information being taught is new, the expository advance organizers would be used. For example, prior to presenting information on how to control a center lathe to a worker that knows nothing about it, an expository advance organizer would first briefly describe the equipment and its function (see Fig. 1-30(a)).

(2) Comparative advance organizers

If people have known similar to the information being presented, comparative advance organizers are useful. For example, a center lathe operator is taught how to control a CNC lathe, a comparative advance organizer would compare and contrast features and operations between the familiar and the new lathe.

### 4. Forgiveness

Human error is inevitable, but it need not be catastrophic. For providing a sense of security and stability, forgiveness could be used in design. Forgiveness helps prevent errors before occurring, and minimizes the negative consequences of occurring errors.

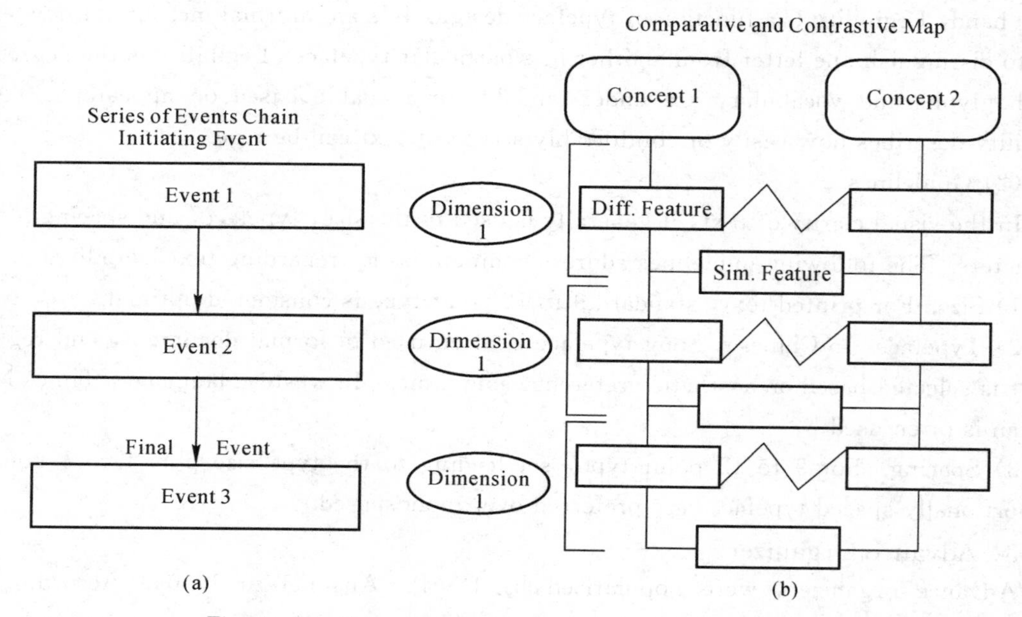


(a) (b)

Fig. 1 - 30 Expository and comparative advance organizers

(a) expository; (b) comparative

There are four kinds of common modes of forgiveness in designs.

(1) Good affordances

Using appropriate physical characteristics of the design can make sure its correct, an example in Fig. 1 - 31.

(2) Reversibility of actions

Performed actions can be reversed if an error occurs or the changing intent, for example, to cancel a floor pressing the floor button it again on elevator or undo function in software (see Fig. 1 - 32).

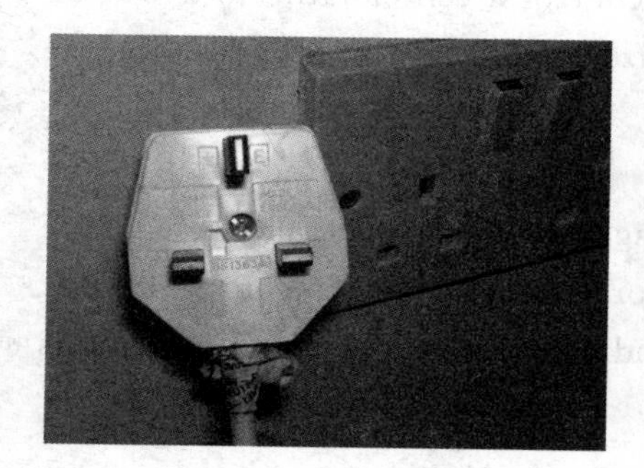

Fig. 1 - 31 Good affordances

Fig. 1 - 32 Redo

(3) Confirmation and warnings

Verification of intent is required before critical actions are allowed by prompt dialog box, signs, or alarms.

(4) Help

Help information is necessary in basic operations, troubleshooting and error recovery.

In designs, effectively use these strategies require minimal confirmations, warnings, and help. If the affordances are good, help is less necessary; if actions are reversible, confirmations and warnings are less necessary. In addition, too many confirmations or warnings impede the flow of interaction.

5. Hierarchy

Hierarchy is an arrangement of items that ranked one above another. A hierarchy is typically depicted as a pyramid, where the height of a level represents that level's status and width of a level represents the quantity of items at that level.

In design, hierarchical organization is the simplest structure for visualizing and understanding complexity, and one of the most effective ways for enhancing the visibility of a system, for example, multi-level software menus.

One way of looking at Biology is to put it in a hierarchy of subjects like so Fig. 1－33. In general, relationships among elements according to hierarchy is primarily their relative left-right and top-down positions, but is also influenced by their proximity, size, and the presence of connecting lines. Superordinate elements are referred to as parent elements, and subordinate elements as child elements.

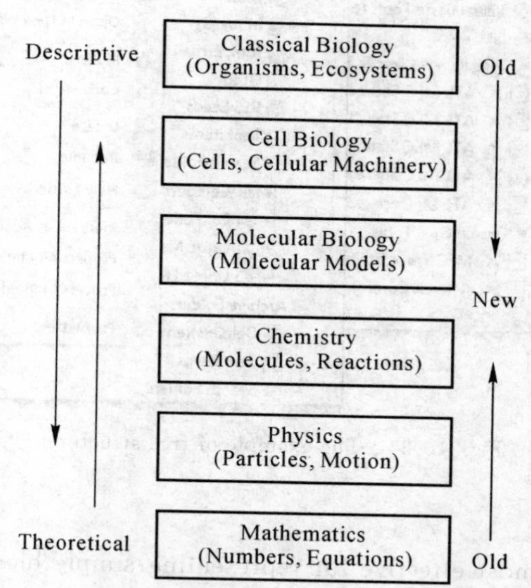


Fig. 1－33　The hierarchy of subjects at Biology

There are two basic ways to visually represent hierarchy: trees and nests.

(1) Tree structures

Representation of tree structures resembles a tree locating child elements below or to the right of parent elements. Tree structure (see Fig. 1－34) has a member that has no superior. This member is called the "root" or "root node", and thought of as the starting node. The lines connecting elements are called "branches", the elements are called "nodes". Nodes without children are called end-nodes. Tree structures are effective model for representing hierarchies of moderate complexity.

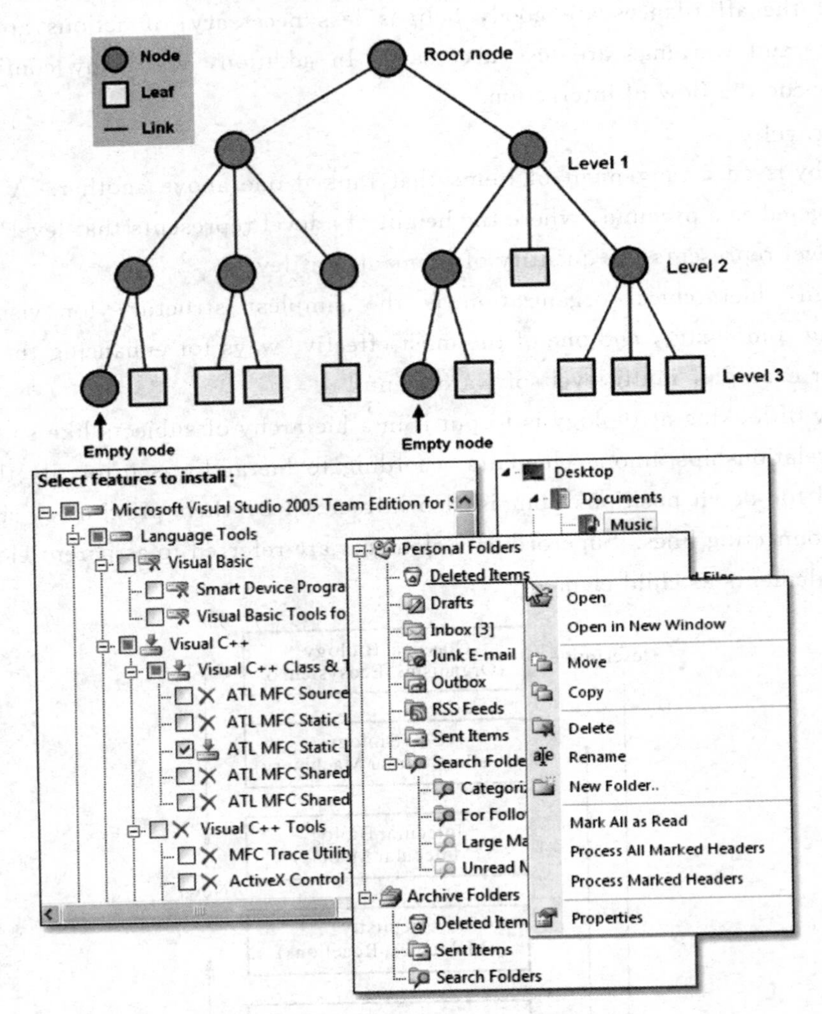


Fig. 1-34 The example of tree structures

(2) Nest structures

Nest structures are most effective for representing simple hierarchies. Nest structures (see Fig. 1-35) illustrate hierarchical relationships by visually containing child elements within parent elements, for example, matryoshka doll. Every element represents a nested hierarchy where each level contains only one object. Nest structures are most commonly used to group information and functions, and to represent simple logical relationships.

6. Accessibility

Accessibility is different from usability, which is the extent to which a product can be used by specified users to achieve specified goals. Accessibility is the degree to which a product, device, service, or environment is available, without modification, to as many people as possible. Accessibility in designs should assert that system is usable for diverse abilities, in special, disabilities.

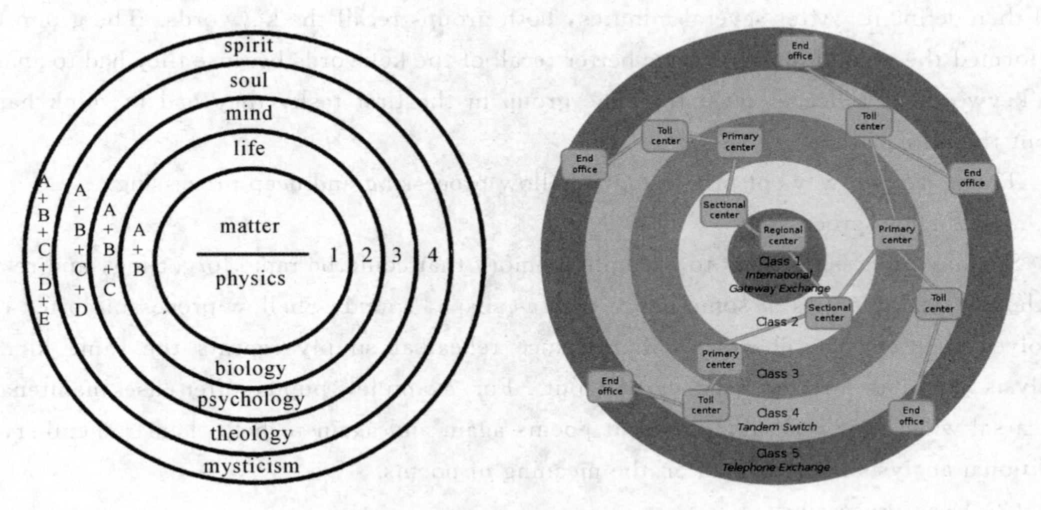


Fig. 1-35　The example of nest structures

Perceptibility and operability are two main characteristics of accessible designs.

(1) Perceptibility

Perceptibility is achieved if anyone can perceive the design, regardless of sensory abilities. In design, it is basic guideline for improving perceptibility that should present information using several methods, such as textual, image, voice, and tactile.

(2) Operability

Operability is achieved if anyone can use the design, regardless of physical abilities. In design, it is basic guideline for improving operability that use good affordances, forgiveness and constraints, and provide diverse position or layout controls.

7. Levels of processing

The levels-of-processing effect is identified by Fergus I. M. Craik and R. S. Lockhart in 1972, who describe memory recall of stimuli as a function of the depth of mental processing. The deeper the level of processing, the easier the information is to recall (see Fig. 1-36). Depth of processing is a phenomenon of memory in which information that is analyzed deeply is better recalled than information that is analyzed superficially.

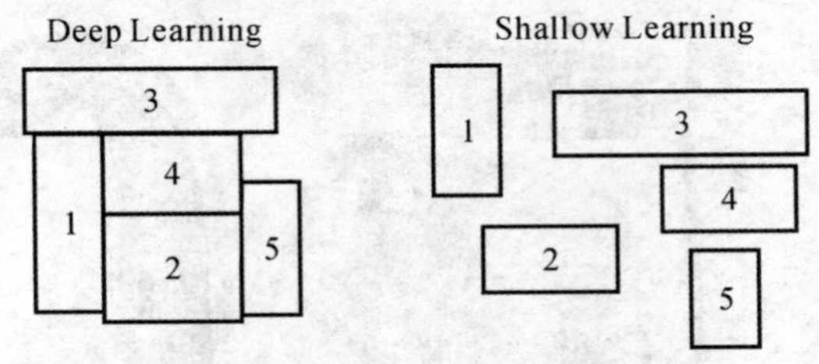


Fig. 1-36　Deep learning and shallow learning

For example, there are two groups to achieve two tasks that involve interacting with and recalling the same information. In the first task, a group of people locate a keyword in a list and circle it. In the second task, another group of people locate a keyword in a list, circle it, and then define it. After several minutes, both groups recall the keywords. The group that performed the second task will have better recall of the keywords because they had to analyze the keywords at a deeper level than the group in the first task; they had to think harder about the information.

There are two ways of processing, shallow processing and deep processing.

(1) Shallow processing

Shallow processing leads to a fragile memory that could be rapid forgetting, and relates to the physical qualities of something, such as shape, sound. Shallow processing often only involves maintenance rehearsal. Maintenance rehearsal simply repeats the same kind of analysis that has already been carried out. For example, pupils often use maintenance rehearsal when they read aloud ancient poems again and again to help them remember, no additional analysis is performed on the meaning of poems.

(2) Deep processing

Deep processing results in a more durable memory, and relates to connotation of something. Deep processing often involves elaboration rehearsal. Elaborative rehearsal involves a deeper, more meaningful analysis of the information. For example, middle school students engage in elaborative rehearsal when they read a ancient poems and then have to analysis as to word and sentence meaning, and conclude the writer's thought. Generally, elaborative rehearsal results in recall performance that is two to three times better than maintenance rehearsal.

## 1.3 实用性设计准则 Usability principles

1. Aesthetic usability effect

Aesthetic is a philosophical theory of art, and relating to a sense of the beautiful (see Fig. 1 - 37). The word aesthetic is derived from a Greek word meaning "esthetic, sensitive". The term aesthetic is used by German philosophers for a theory of the beautiful in 1734. Usability is the ease of use, and achieves specified goals with effectiveness, efficiency, and satisfaction in a specified context of use.

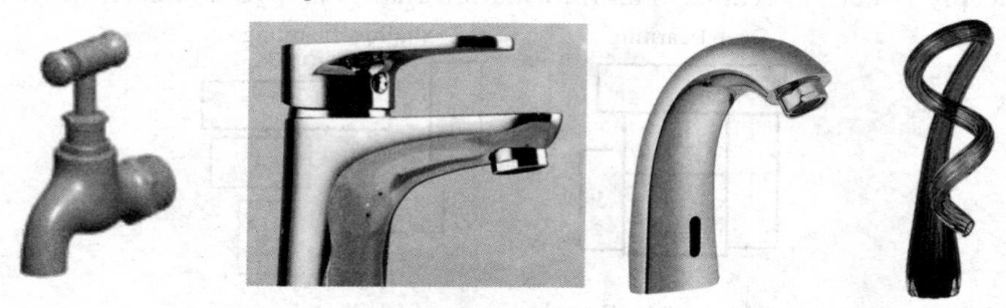

Fig. 1 - 37 Which faucet is aesthetic and which faucet is easy to use?

The aesthetic usability effect describes that aesthetic design influence on the user's sense using product, which people perceive more aesthetic designs as easier to use than less aesthetic designs. This perception is encouraged whether or not the design is actually easier to use, but it is an important part of the user experience whether they are or not. The effect has been observed in many experiments, and has significant implications for the acceptance, use, and performance of the design. Aesthetic designs not only look easier to use, and also have a higher probability of being used, whether or not they actually are easier to use. So even if something is designed in a very user-friendly way, if it is not aesthetically pleasing it is less likely to be used, making it's usability of less value. Then there is a similar phenomenon, humans who are more beautiful tend to attract more attentions and be treated better.

Aesthetics play an important role in the way a design is used. Therefore it is vital to foster a positive attitude to a product by using aesthetic designs to encourage positive feelings such as affection, loyalty and patience, which the perceptions should promote creative thinking, problem solving, tolerate more faults of product, improve the chances of making a sale and used over time.

Conversely, less aesthetic designs should lead to negative feelings and tend to narrow thinking, stifle creativity, reduce the desire to possess the product and never use it.

Thus it is important to create aesthetic designs for product. Good design means that beauty and usability are in balance. An object that is beautiful to the core is no better than one that is only pretty if they both lack usability.

## 2. 80/20 Rule

The 80/20 Rule is also called the Pareto Principle (see Fig. 1-38). In 1906, Italian economist Vilfredo Pareto investigated the unequal distribution of wealth in his country, observing that 20 percent of the people owned 80 percent of the wealth. In the late 1940s, Dr. J. M. Juran inaccurately attributed the 80/20 Rule to Pareto, calling it Pareto Principle.

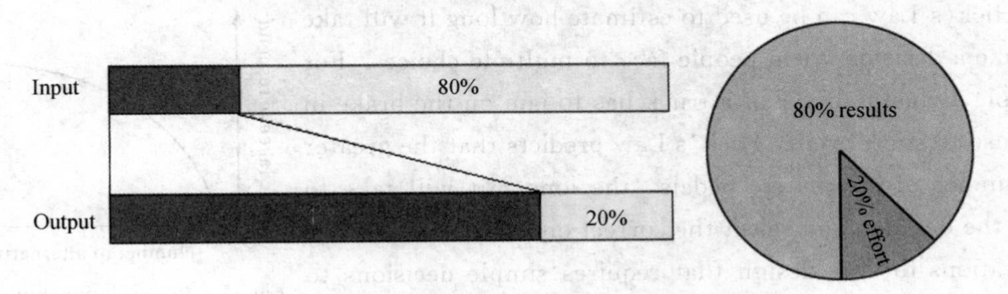


Fig. 1-38 The relationship of 80 and 20

The 80/20 Rule is a high percentage of effects in any large system are caused by a low percentage of variables. The 80/20 rule can be observed in most of large systems, including in economics, management, business, engineering, software, health, and so on. A few examples of the 80/20 rule include:

80% of profits come from 20% of customers;
80% of product's usage comes from 20% of its features;
80% of sales come from 20% of products;
80% of complaints come from 20% of customers;
80% of progress comes from 20% of the effort;
80% of errors are caused by 20% of the components;
80% of the program resources are used by 20% of the codes of program;
80% of the injuries are used by 20% of the program;
80% of health care resources are used by 20% of patients.

On the other hand, 80/20 rule could potentially change life or decision. For example, there are two main factors about selling a house, price and effective marketing. 80% is how this house is priced, and 20% is effectively marketing property to buyers and other real estate agents. Because this house is not the only one on the market, pricing it comparably to the market is vital to getting it sold. Therefore the price of this house plays the biggest role in the buyer's decision.

The 80/20 rule is useful for realizing greater efficiencies in design. For example, because the 20 percent of a product's features are used 80 percent of the time, design and testing resources should focus primarily on those features. The remaining 80 percent of the features should be reevaluated to verify their value in the design.

3. Hick's Law

Hick's Law, or the Hick-Hyman law was examined in the early 1950s by British psychologist Hick and Hyman. And Hick's Law is a model of human-computer interaction that describes the time it takes for a person to make a decision as a result of the possible choices he or she has. Then the time it takes to make a decision increases as the number of alternatives increases. Hick's Law states that the time required to make a decision is a function of the number of available options (see Fig. 1 - 39).

Hick's Law can be used to estimate how long it will take to make a decision when people face to multiple choices. For example, when a driver of a truck has to jam on the brake in response to some event, Hick's Law predicts that the greater the number of alternative pedals, the longer it will take to make the decision and select the correct one. Hick's Law has implications for the design that requires simple decisions to be made based on multiple options.

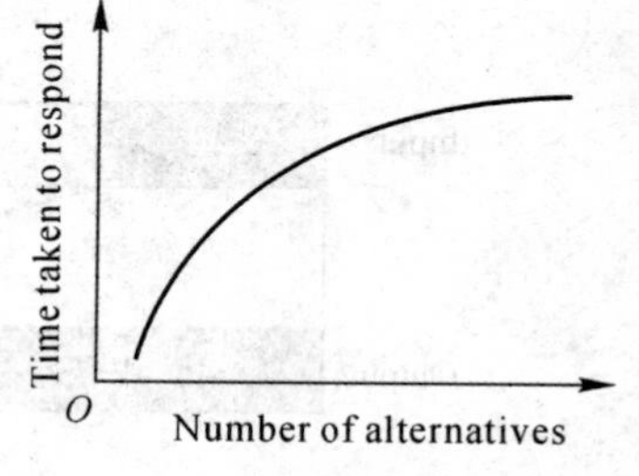

Fig. 1 - 39 Relationship of time and choices

There are four basic steps for achieving a task, which are to ascertain a problem or goal, to assess the available options to solve the problem or achieve the goal, to select an option, and implement the option. Hick's Law applies to the third step: decide on an option.

Hick's Law is more applicable to tasks making simple decision in which there is only one response to each stimulus, that is 1 happens push button 1, then 2 happens push button 2. The efficiency of a design can be improved by application of Hick's Law. But Hick's Law does not apply to complex projects of options.

In addition, it is known that the stimulus and response compatibility affect the decision reaction time for the Hick's Law. This is that the response should be similar to the stimulus. For example, turning a wheel to turn the wheels of the car is good stimulus and response compatibility. The action the user performs is similar to the response the driver receives from the car.

If a design that involve decisions based on a set of options should use Hick's Law. And designing time-critical tasks, minimize the number of options can reduce response times and minimize errors.

## 4. Cost-Benefit Analysis

Cost Benefit Analysis (CBA) is also called Benefit-Cost Analysis (BCA). Cost benefit analysis is a systematic process to calculate and compare benefits and costs of a project. Although a cost benefit analysis is most commonly done on financial questions, it can be used for almost anything.

The results of cost benefit analysis decide whether the project will be performed or not. An activity will be pursued only if its benefits are equal to or greater than the costs in Fig. 1-40. If the costs associated with interacting with a design outweigh the benefits, the design is poor. If the benefits outweigh the costs, the design is good.

Fig. 1-40 Comparison of benefits and costs

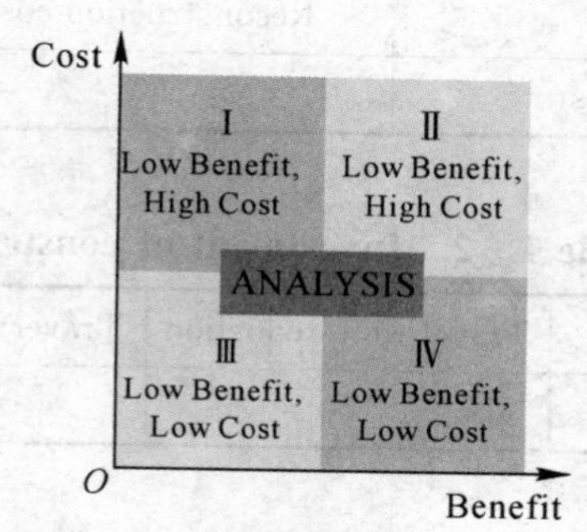


Fig. 1-41 Cost-benefit analysis

The quality of every design aspect can be measured using the cost-benefit. How much reading is too much to get the point of a message? How long is too long for a person to wait for a Web page to download? The answer to all of these questions is that it depends on the benefits of the interaction to that specific user.

Once the results of cost benefit analysis are quantified, the resultant data can be charted on the Cost Benefit model in Fig. 1-41 and can be used to ascertain quickly the cost-benefit results of a program. There are four regions. In region Ⅰ, the programs are described as low benefit, high cost so that they become the least desirable programs. For example, walking some distance to see performance is a cost. The pleasure of seeing interesting things constitutes a benefit. In region II, the programs are high benefit, high cost. High benefit makes them desirable, but the high cost factor prevents their widespread use. And if the above performance is extremely attractive, some people could go to see performance even though it takes a lot of time to walk. In region III, the programs are described as low benefit, low cost and they are "trouble" programs because low cost are attractive. Whereas they do not produce a desirable level of benefits and thus should be examined to see if there are better alternatives. And if the performance is very boring, but it takes a little or no time to walk, many people could go to see performance because they have not do other more interesting thing and the cost is low. In region IV, the programs are high benefit, low cost and are the ideal results. If some programs fall near the intersection of all the regions, cost-benefit analysis alone is inconclusive for determination of implied action.

For example, construction of a highway is feasible by the cost benefit analysis. The expenditure invested in facilities is cost (see Table 1 - 1). They include in reconstruction and maintenance cost. And service of the facilities is benefits (see Table 1 - 2). After the highway will be available, travel time will be reduced, travel cost will also be decreased, and the social cost will be reduce due to reducing traffic accident.

Then total of benefit divided total of cost is the cost and benefit ratio. The cost and benefit ratio in construction of a highway is 2. 5. And this project is high benefit and low cost in region IV.

**Table 1 - 1　Investment of construction of a highway (unit: ten thousands yuan)**

| | Reconstruction cost | Maintenance cost | Total |
|---|---|---|---|
| Cost | 50 | 10 | 60 |

**Table 1 - 2　Investment of construction of a highway (unit: ten thousands yuan)**

| | Travel time reduction | Travel cost reduction | Traffic accident reduction | Total |
|---|---|---|---|---|
| Benefit | 100 | 20 | 30 | 150 |

In design, it is necessary to consider the cost-benefit principle in all aspects of design.

5. Entry point

In trade, the entry point is usually a component of a predetermined trading strategy for minimizing investment risk and removing the emotion from trading decisions. Recognizing a good entry point is the first step in achieving a successful trade. In computer programming, an entry point is a memory address, corresponding to a point in the code of a computer program which is intended as the destination of a long jump, be it internal or external. Therefore entry point is a point of physical or attentional entry into a design.

The different design has different point. For example, the entry point of a book is its cover, that of an internet site is its first page, and that of building is its lobby. This initial impression of a system or environment is largely formed at the entry point to a system or environment. And this impression greatly influences subsequent perceptions and attitudes, which then affects the quality of subsequent interactions. For example, entering a building opening a black door squeak, followed by a dim corridor, followed by a narrow and small lobby with several closed doors, all this to enter a building that may or may not have the place the person was looking for. Such problems in entry point design annoy and deter visitors. Either way, it does not promote additional interaction.

There are three key elements to improve those problems of entry point design. They are minimal barriers, front area, and vectoring.

(1) Minimal barriers

A barrier is a physical structure which blocks or impedes something, which is perceived through vision, audition and tactus. A good entry point should reduce barriers to be minimal as possible. For example, use of glass minimizes visual barriers in the Apple retail store in Fig. 1-42.

But sometimes barriers should not encumber entry points. For example, traffic barriers are used to keep vehicles within their roadway, and a sound barrier is an exterior structure designed to protect inhabitants of sensitive land use areas from noise pollution, and salespeople standing at the doors of retail stores, and so on. Barriers can also be aesthetic such as curtain, landscape, etc.

Fig. 1-42 Use of glass minimizes visual barriers in the Apple retail store

(2) Front area

Entry points should allow people to become oriented and clearly survey available options. Front area include cover of book that provide some information of contents, store entrances that provide a clear view of store layout and aisle signs (see Fig. 1-43), or Internet pages that provide good orientation cues and navigation options. Then front area should provide sufficient time and space for a person to review options with minimal distraction or disruption (see Fig. 1-44 and Fig. 1-45).

(3) Vectoring

Vectoring should attract and pull people through the entry point. Vectoring get people to incrementally approach, enter, and move through the entry point. Modes of vectoring can be compelling headlines from the front page of a newspaper, greeters at restaurants, or the display of products just beyond the entry point of a store.

Effectiveness of the entry point in a design could be enhanced by reducing barriers, establishing clear front, and using vectoring.

Fig. 1-43 Store layout and signs

Fig. 1-44 Products line the periphery of the space in the Apple retail store

Fig. 1-45 Large glass staircase acts as a secondary lure in the Apple retail store

6. Wayfinding

Wayfinding is a techniques used by travelers over land and sea to find relatively unmarked and often mislabeled routes. Wayfinding also refers to the organization and communication of our dynamic relationship to space and the environment. As a whole, wayfinding is the process of using spatial and environmental information to navigate to a destination, such as a park, the wilds of a forest, or an internet site.

There are four basic wayfinding elements: orientation, desire path, markers, and destination.

(1) Orientation

Orientation is the act of orienting or a relative position, and refers to determining one's location relative to nearby objects and the destination. Using landmarks and signage can improve orientation. Landmarks provide locations with easily remembered identities (see Fig. 1-46). Signage can tell people where they are and where they can go. Then good signage can be interpret and achieve quickly the goal. For example, people can more quickly leave a building according to the signage in Fig. 1-47 than in Fig. 1-48.

Fig. 1-46　Landmarks

Fig. 1-47　Good signage

Fig. 1-48　Bad signage

(2) Desire path

Path refers to routes to get to the destination. Desire path usually represents the shortest or most easily navigated route between an origin and destination. It can increase efficiency to reduce the number of choices, and use signs or prompts at decision points.

(3) Markers

In wayfinding, a marker is an object that marks a locality. According to the chosen route, markers should be set up at focal points such as beginnings, middles, and ends, and at places that correspond to intersections in order to confirm that it is leading to the destination.

(4) Destination

This graphic information is provided at the point of destination so as to recognize the destination. Typically it includes building signage, barriers, and location identifiers.

7. Constraint

Constraint is the state of being restricted, or compelled to avoid some actions that could be performed on a system. For example, hiding options that are not available at a particular time effectively constrains the options that can be selected.

Constraint also represses one's own feelings or behavior.

There are physical constraints and psychological constraints.

(1) Physical constraints

Physical constraints limit the range of possible actions by redirecting physical motion in specific ways. Physical constraints are useful for reducing the sensitivity of controls to unwanted inputs, and denying certain kinds of inputs altogether.

There are two methods of physical constraints: limit value and substance.

Limit value can set maximum or minimum to constrain the motion. Channels or grooves are often used into linear or curvilinear motion, and are useful in situations where the control variable range is relatively small and bounded. For example, the scroll bar of volume can change the volume of audio player in Fig. 1-49. And rotary knob is often used into rotary motion, and is useful in situations where control real estate is limited, or the control variables are very large or unbounded(see Fig. 1-50).

Substance can be achieved to set physical structure which blocks or impedes something, such as barriers or edges, and useful for denying errant or undesired actions (see Fig. 1-51).

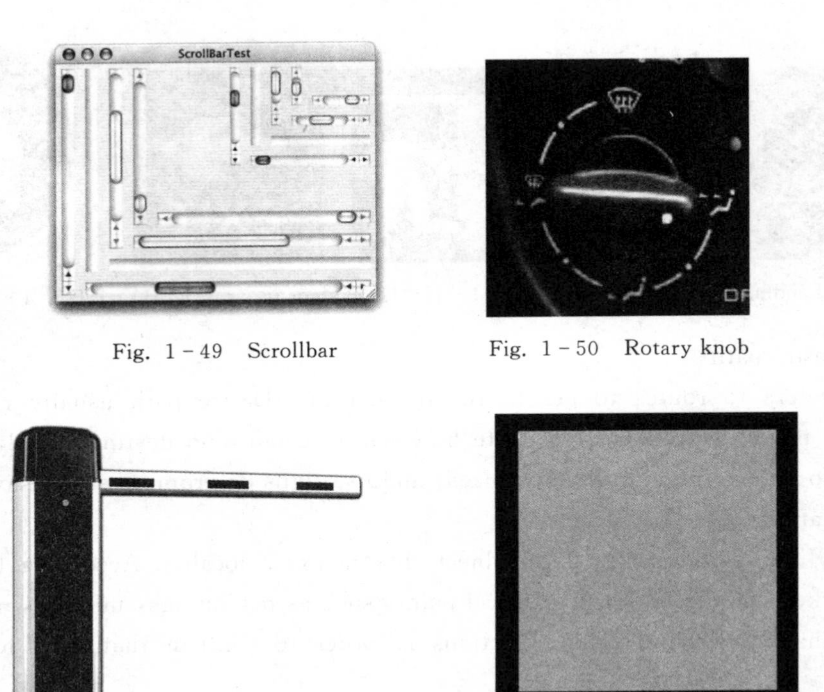


Fig. 1 - 49  Scrollbar

Fig. 1 - 50  Rotary knob

Fig. 1 - 51  Barrier and edge

(2) Psychological constraints

Psychological constraints limit the range of possible actions by leveraging the way people perceive. There are two kinds of psychological constraints: symbols and conventions.

Symbols are some simple signs of text or icon. Symbols influence behavior of people by communicating meaning through language. Symbols are useful for labeling, explaining, and warning using visual, aural, or tactile representation.

Conventions influence behavior of people based on learned traditions and practices. For example, red means stop, green means go. Conventions indicate common methods of understanding and interacting, and are useful for making systems consistent and easy to use (see Fig. 1 - 52).

Fig. 1 - 52  Signs and traffic light

Proper application of constraints makes designs easier to use and dramatically reduces the probability of error during interaction.

8. Error

Error is a deviation from accuracy or correctness, so is also an action or omission of action yielding an unintended result.

Some accidents are caused by human operation, yet some accidents are actually due to design errors. Proper understanding of the causes of errors can greatly reduce their frequency and severity.

There are two basic types of error: slips and mistakes.

(1) Slips

Slips are the result of automatic, unconscious processes, and refer to as errors of action or errors of execution. Slips occur when an action is not what was intended. For example, a slip occurs when a person dials a wrong phone number frequently dialed because he is interrupted by knocking door.

There are two types of slips: action and attention.

1) Action. Slip occurs when repetitive tasks or habits has changes. Then it could be avoided by providing clear and distinctive feedback, using confirmations for critical tasks, or considering constraints, affordances, and mappings. For example, confirmations are useful for disrupting behaviors and verifying intent.

2) Attention. Slip occurs when distractions and interruptions appear. Then it could be avoided by provide clear orientation and status, using highlighting to focus attention, or using alarms to attract attention for critical situations. For example, clear orientation and status cues are useful for enabling the easy resumption of interrupted procedures (see Fig. 1-53).

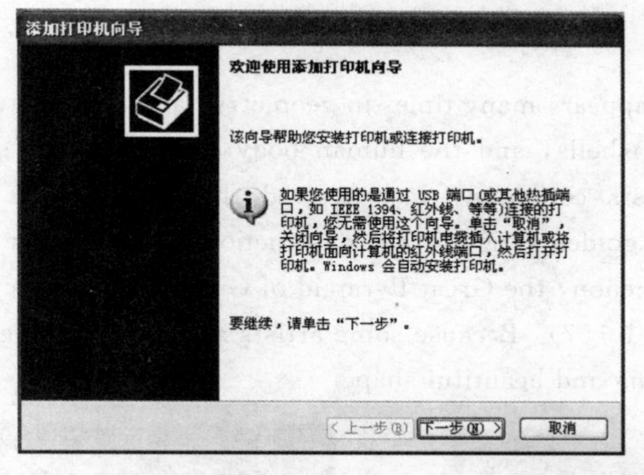


Fig. 1-53　Confirmation, constraint and status cue

(2) Mistakes

Mistakes are an erroneous belief, and caused by conscious mental processes, and frequently result from stress or decision-making biases. Mistakes occur when an intention is inappropriate.

There are three types of mistakes: perception, decision, and knowledge.

1) Perception. Mistake is caused by incomplete or ambiguous feedback. It could be avoided by improving situational awareness, providing clear and distinctive feedback, or tracking and displaying historical system behaviors. For example, historical displays are useful for revealing trends that are not detectable in point-in-time displays.

2) Decision. Mistake is caused by stress, decision biases, and overconfidence. It could be avoided by minimizing information and environmental noise, using checklists and decision trees, and training on error recovery and troubleshooting. For example, decision trees and checklists are useful decision-making and troubleshooting tools, especially in times of stress.

3) Knowledge. Mistake is caused by lack of knowledge and poor communication. It could be avoided by using memory and decision aids, standardizing naming and operational conventions, or training using case studies and simulations. For example, memory mnemonics are useful strategies for remembering critical information in emergency situations.

## 1.4 渲染性设计准则 Attractable principles

1. Golden ratio

The golden ratio is a special number approximately equal to 1.618 (see Fig. 1-54). This ratio was named the golden ratio by the Greeks, which the whole length of line (*AB*) divided by the longer part (*AC*) or the longer part (*AC*) divided by the smaller part (*BC*). Golden ratio is also a ratio within the elements of a form, such as height to width, approximating 1.618.

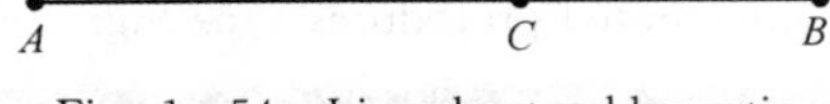


Fig. 1-54 Line about golden ratio

The golden ratio appears many times in geometry, art, architecture and other areas. In nature, pinecones, seashells, and the human body all exhibit the golden ratio (see Fig. 1-55). In art, artists commonly incorporated the golden ratio into their paintings. Stradivari utilized the golden ratio in the construction of his violins (see Fig. 1-56). In architecture, the Parthenon, the Great Pyramid of Giza, and Stonehenge also all exhibit the golden ratio (see Fig. 1-57). Because some artists and architects believe the Golden Ratio makes the most pleasing and beautiful shape.

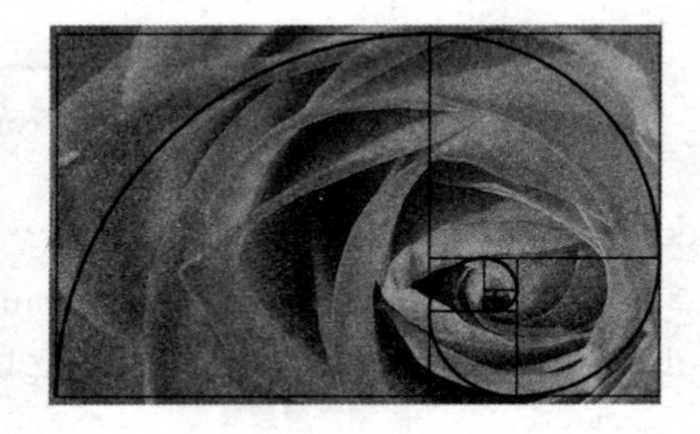

Fig. 1-55 Examples of the golden ratio in nature

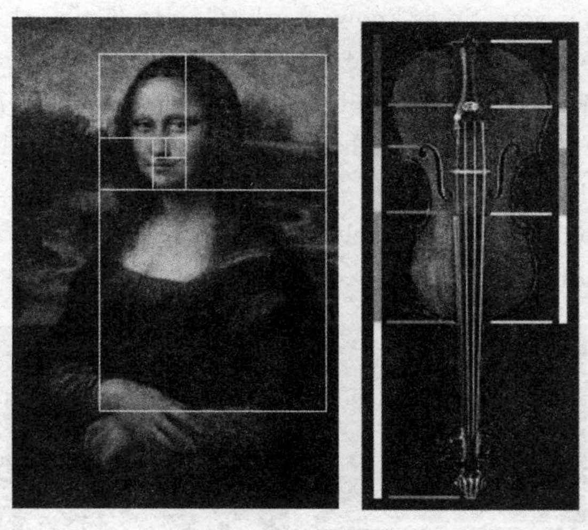

Fig. 1－56　Examples of the golden ratio in art

Fig. 1－57　Examples of the golden ratio in architecture

The golden ratio continued influence on design. Always consider the golden ratio in design.

2. Rule of Thirds

Rule of Thirds is a principle of photographic composition derived from the use of early grid systems. Subjects or regions of an image are composed along imaginary lines which divide the image into thirds both vertically and horizontally, creating an invisible grid of nine rectangles and four intersections (see Fig. 1－58). Important elements of the composition should be placed at these intersections. Then the asymmetry of composition is generally agreed to be aesthetic.

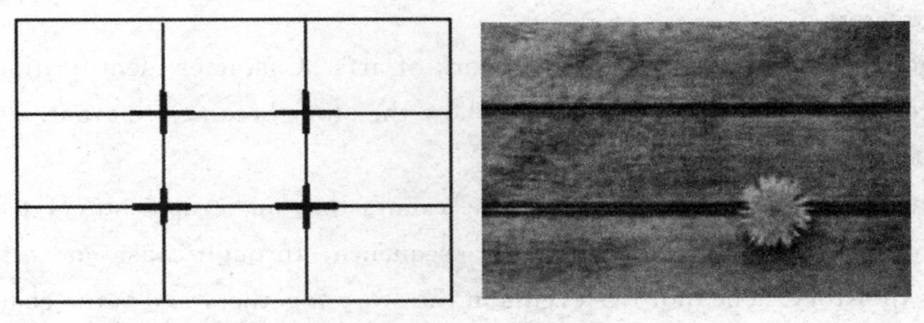

Fig. 1－58　Grid of rectangles and intersections

The Rule of Thirds is sometimes known as the Golden Ratio, although dividing a design into thirds yields a ratio (1.5) different from the golden ratio (1.618). The users of the technique may have decided that the simplicity of its application compensated for its rough approximation. Using the Rule of Thirds helps produce nicely balanced easy on the eye pictures (see Fig. 1-59).

The rule of thirds generally works well, is easy to apply, and should be considered when composing elements of a design.

Fig. 1-59 Images of rule of thirds

3. Storytelling

Storytelling predates writing, with the earliest forms of storytelling primarily oral combined with gestures and expressions. It is a method of creating imagery, emotions, and understanding of events using words, images and sounds.

Storytelling is uniquely human. Storytelling can be oral, visual, textual, or digital media. A storyteller can be any instrument of information presentation that engages an audience to experience a set of events. Telling a story is the most powerful way to activate people's brains. Everyone all enjoy a good story, whether it's a novel, a movie, or simply something one of friends is explaining to him. Therefore stories or narratives have been shared in every culture as a means of entertainment, education, and business or advertisement.

Good storytelling experiences generally require certain fundamental elements. Crucial elements of stories and storytelling include setting, characters, plot, and point of view.

(1) Setting

The setting orients the audience, providing a sense of time and place for the story.

(2) Characters

A character is a person in a narrative work of arts. Character identification is how the audience becomes involved in the story, and how the story becomes relevant.

(3) Plot

The plot is a literary term defined as the events that make up a story, particularly as they relate to one another in a pattern, in a sequence, through cause and effect how the reader views the story. The plot ties events in the story together, and is the channel through which the story can flow.

(4) Point of view

Point of view is the set of methods the author of a literary, theatrical, cinematic, or musical story uses to convey the plot to the audience. Music, lighting, and style of prose create the emotional tone of the story.

4. Archetypes

Carl Gustav Jung first used the term Archetype in 1919. In addition to the personal unconscious, he posited a collective unconscious which is formed of two components, the Instincts and the Archetypes. The Instincts are innate capability or aptitude, and are impulses which carry out actions from necessity, and have a biological quality similar to the homing instincts in bird. Instincts determine actions. Yet in the same manner, Carl Jung suggests that there are innate unconscious modes of understanding which regulate our perception itself. These are the Archetypes that the original pattern or model from which all things of the same kind are copied. They are inborn forms of "intuition".

Archetypes have no material existence and reveal themselves only as images. For example in all ages and cultures mankind imagined itself in communion with a "wise spirit". One of the most common forms for this conception is the image of the wise old man found in innumerable myths and legends. Other Archetypes could be the experienced hero, the beautiful princess, the clever old queen, the witch, the nurturing parent, or the loyal friend.

An archetype is a universally symbol, statement, or pattern of behavior and often used in myths and storytelling across different cultures. At some time, an archetype is a model of a person, personality, or behavior. As a whole, Archetypes are universal patterns of theme and form resulting from innate biases or dispositions.

Archetypes are found in many fields and are believed to be a product of unconscious biases. Identifying and aligning appropriate archetypes in design will reinforce its attraction.

For example, Harley-Davidson often abbreviated H-D or Harley, is an American motorcycle manufacturer. It designs its product and brand with the outlaw archetype, emphasizing freedom and living outside the rules of society (see Fig. 1 - 60). Harley-Davidson's motorcycles have black and chrome features and distinctive sound of engine.

Fig. 1 - 60 Product and brand of Harley-Davidson

And an example, Nike is an American multinational corporation that is engaged in the design, development and worldwide marketing and selling of footwear, apparel, equipment, and accessories. Nike is known as the winged Goddess of victory in Greek mythology, so the Hero archetype is designed in its product and brand. Michael Jordan, Liu Xiang, and Kobe Bryant are all shown wearing Nike products in advertisement (see Fig. 1-61).

Fig. 1-61 Advertisements of Nike

In storytelling, archetypal themes are all too familiar. For example, there are a number of animated films (see Fig. 1-62).

Fig. 1-62 Characters of animated films

Archetypes influence perception on an unconscious, it could increase attraction using archetypal themes and forms in design.

5. Attractiveness Bias

Attractiveness Bias is a psychological term, refer to a tendency to see attractive people as more intelligent, successful, moral, and sociable than unattractive people. In other words, it is the tendency to assume that people who are physically attractive also possess other socially desirable personality traits. Physical attractiveness refer to some physical features are attractive in both men and women, particularly bodily and facial symmetry, clarity and smoothness of skin, and vivid color of eyes and hair, etc.

Attractive people are generally perceived more positive, health and fertility than unattractive people. They receive more attention from the opposite sex, receive more affection from their parents, and receive more supports from the colleagues than do unattractive people. For example, there are considerable cases that physical attractiveness affects directly employment decision making, which the more attractive an individual has the greater the likelihood to be hired.

The attractiveness bias is a function of both biological and environmental factors. Using the attractiveness bias in design can reinforce the attraction of product, such as marketing and advertising with images of attractive people (see Fig. 1 - 63).

Fig. 1 - 63　The advertisement of TCL

6. Baby-Face Bias

Baby-Face referring to some things or people with round features, large eyes, small noses, high foreheads, short chins, lighter skin and curly hair are perceived as babylike, such as teletubbies, Calabash Brothers (see Fig. 1 - 64). And animals with "baby faces" are also extremely attractive so that people think puppies, and kittens so cute.

Baby-Face Bias is a tendency to perceive people and things with baby-faced features as more naive, helpless, and honest.

Fig. 1 - 64　Teletubbies and Calabash Brothers

In life, babies with weak baby-face features could receive less positive attention from adults. Baby-faced adults are subject to a similar bias. Baby-faced adults could receive more tolerance and reliance, but they are perceived as simple and naive, and have difficulty being taken seriously in situations where expertise or confrontation is required.

In design of children product, cartoon characters could attract more attention of children. In marketing and advertising, mature-faced people can convey expertise and authority (see Fig. 1 - 65).

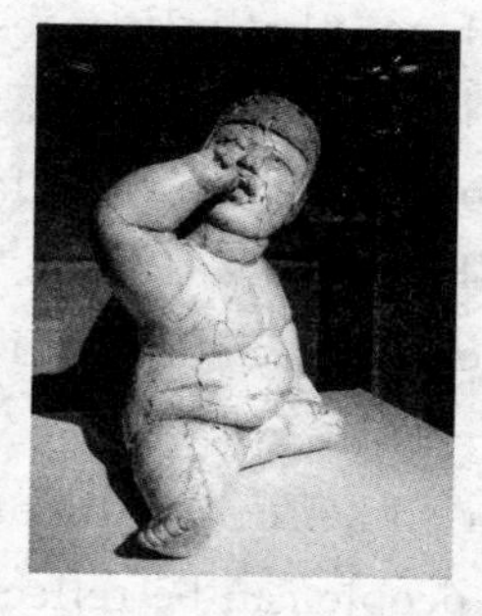

Fig. 1 - 65　Photo of a famous "baby-face" artifact from Las Bocas

### 7. Face-ism Ratio

The term "face-ism" was initially defined by Archer, Iritani, Kimes and Barrios in their paper "five studies of sex differences in facial prominence" in 1983. The facial prominence was measured by a face-ism index, which is the ratio of two linear measurements, with the distance in a depiction from the top of the head to the lowest visible point of the chin being the numerator and the distance from the top of the head to the lowest visible part of the subject's body the denominator. In Fig. 1-66, Face-ism ratio is Length of face ("F") divided by total length of body ("B"). An image without a face would have a face-ism ratio of 0.00, and an image with only a face would have a face-ism ratio of 1.00. They found that facial prominence of men has been much higher than that of women. Women are often portrayed in media with an emphasis on their bodies, while men are depicted with an emphasis on their faces (see Fig. 1-67).

Fig. 1-66 Ratio of face

Fig. 1-67 Examples of face-ism

The ratio of face to body in an image that influences the way the person in the image is perceived. Persons in image with a high face-ism ratio focus attention on their intellectual and personality attributes. Persons in image with a low face-ism ratio focus attention on their physical and sensual attributes of the person. People assess individuals in high face-ism images as being more intelligent, dominant, and ambitious than individuals in low face-ism images (see Fig. 1-68).

Face-ism ratio can help designer to select photos of people in designs. In photographs and drawings, the representation of people can be reflected appropriately by face-ism ratio.

Fig. 1 - 68　Lower ratio, medium ratio and high ratio

8. Defensible space

Defensible space is defined in Newman's book "Design Guidelines for Creating Defensible Space" in 1973. It is "a residential environment whose physical characteristics—building layout and site plan—function to allow inhabitants themselves to become key agents in ensuring their security." Oscar Newman's defensible space theory is largely popular in city design.

Defensible Space (see Fig. 1 - 69) is the buffer creating between a building on property and the grass, trees, shrubs, or any area that surround it. There are territorial markers, surveillance, and clear indications of activity and ownership. Defensible spaces are used to deter crime.

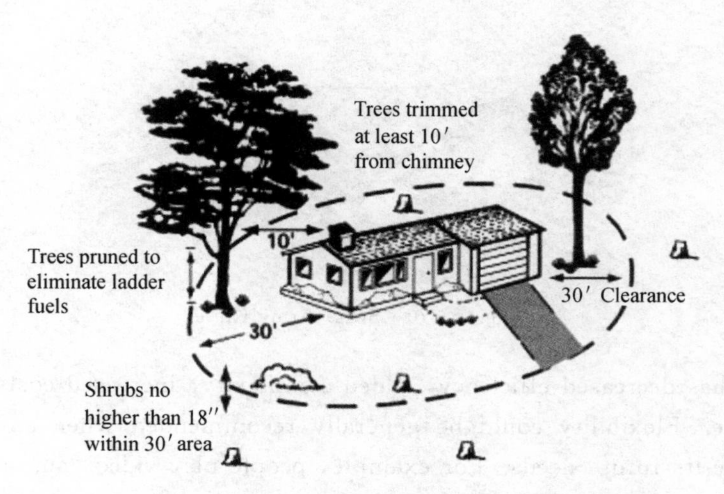


Fig. 1 - 69　The defensible spaces

Territoriality is the establishment of clearly defined spaces of ownership using markers and gates, or visible boundaries such as walls, hedges, and fences. Surveillance is the monitoring of the environment during normal daily activities using external lighting, and walkways.

## 1.5 设计决策 Design decisions

1. Flexibility-usability tradeoff

Flexibility is used as an attribute of various types of systems. In the field of design, it refers to designs that can adapt when external changes occur. As the flexibility of a system increases, its usability decreases. The flexibility-usability tradeoff is to weight the relative importance of flexibility versus usability. Then designers should be advised to consider how well the needs of the user are understood. If user needs are well understood, designers should bias towards simple less-flexible systems. Otherwise, designers should create flexible designs that support multiple future applications.

Flexible design can perform many functions, and can satisfy multiple requirements, but the functions should be less efficiently. Flexible designs are more complex than inflexible designs, and are generally more difficult to use. For example, a Swiss Army Knife, a typical flexible design, is a kind of pocket knife. It generally has many tools, such as a blade, screwdrivers reamer, can-opener, and grips, etc. Some tools are less usable and efficient than corresponding individual tools, but taken together provide a flexibility of use not available from any single tool (see Fig. 1 - 70).

Fig. 1 - 70 Swiss Army Knife

Flexibility has decreased efficiency, added complexity, increased costs of development and manufacture. Flexibility could be generally recommended when an audience cannot clearly anticipate its future needs. For example, people play video game using video game player, personal computer, or tablet computer (see Fig. 1 - 71). A tablet computer is a one-piece mobile computer. It has basic functions of personal computer and mobile phone. It is more flexible devices than personal computers. Data can be input by a hideable virtual keyboard so that some functions are difficult to use, such as word processing, and image processing. And personal computer is also strength as a multipurpose device that can be used in many different ways. Therefore people purchase video game players to play games.

It is a common design mistake to assume that designs should always be made as flexible as possible. When a user has not clear understanding of needs, favor flexible designs to address the broadest possible set of future applications.

Fig. 1－71　Game player，personal computer and tablet computer

2. Feedback loop

Feedback is a process in which information about the past or the present influences the same phenomenon in the present or future. As part of a chain of cause-and-effect that forms a circuit or loop，the event is called "feedback loop". That is the consequences of an event feed back into the system as input and modify the event in the future. All real systems are compose of many such interacting feedback loops，such as animals，machines，businesses，and ecosystems，etc.

There are two types of feedback loops：positive and negative.

(1) Positive feedback

A positive feedback is a process in which an initial change will bring about an additional change in the same direction. For example，renovation of football helmets is a positive feedback (see Fig. 1－72). In order to reduce head and neck injuries in football，designers used plastic football helmets with internal padding to replace leather helmets. The helmets provided more protection，but more head and neck injuries occurred than before. It is main reason that players used their head and neck in increasingly risky ways because the helmet shells are harder. Climate-carbon cycle feedbacks refer to the interaction between temperature change，atmospheric carbon dioxide levels and the carbon cycle (see Fig. 1－73). Global warming might cause loss of carbon from terrestrial ecosystems，leading to an increase of atmospheric carbon dioxide levels.

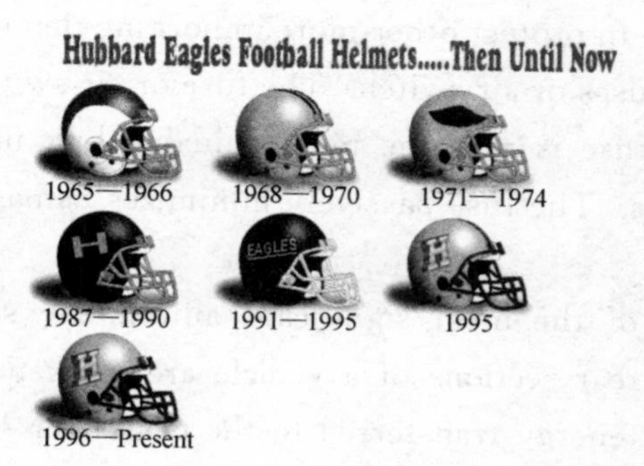


Fig. 1－72　Football helmets

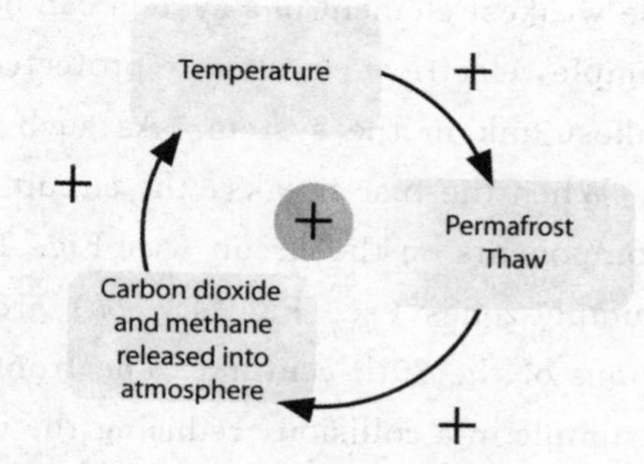


Fig. 1－73　Terrestrial Carbon Cycle Climate Feedback

(2) Negative feedback

Negative feedback occurs when the result of a process influences the operation of the process itself in such a way as to reduce changes. Negative feedback dampens output, stabilizing the system around an equilibrium point.

Negative feedback loops are effective for resisting change. For example, the Segway Human Transporter uses negative feedback loops to maintain equilibrium (see Fig. 1-74). As a rider leans forward or backward, the Segway accelerates or decelerates to keep the system in equilibrium. And in another example, receptors in the carotid arteries detect the change in blood pressure and send a message to the brain. Then the brain will cause the heart to beat slower and thus decrease the blood pressure. Decreasing heart rate has a negative effect on blood pressure (see Fig. 1-75).

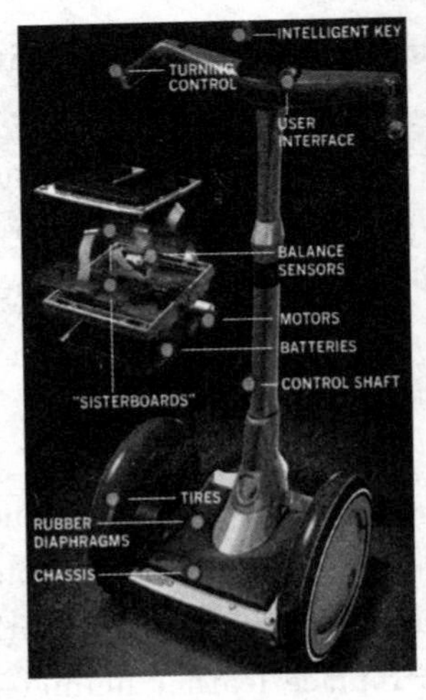

Fig. 1-74 The Segway

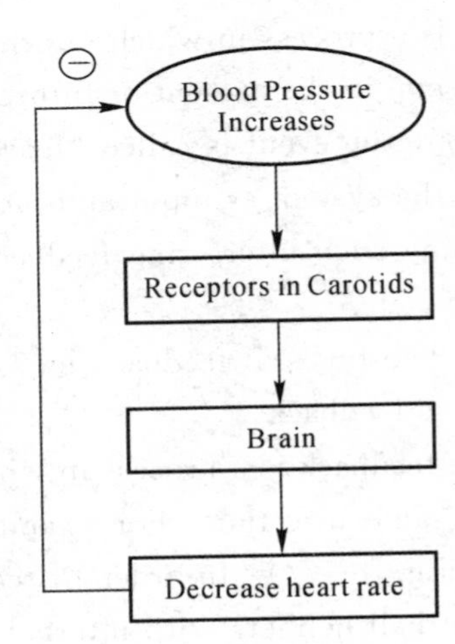

Fig. 1-75 Blood pressure

3. Weakest link

Weakest link is the element that is making the least contribution to the collective achievement of the group. Weakest link is the deliberate use of a weak element that will fail in order to protect other elements in the system from damage. In the other words, a chain is only as strong as its weakest link.

The weakest element in a system can be used to protect other more important elements. For example, electrical circuits are protected by fuses or air switch. The fuse or air switch is the weakest link in the system. As such, the fuse is also the most valuable link in the system. When the fuse breaks, the circuit breaks. The fuse passively minimizes damage to other components on the circuit (see Fig. 1-76).

Crumple zones (see Fig. 1-77) are one of the most significant automobile safety innovations of the 20th century. The front and rear sections of a vehicle are weakened to easily crumple in a collision, reducing the impact energy transferred to the passenger shell. The passenger shell is reinforced to better protect occupants. The total system is designed to sacrifice less important elements for the most important element in the system — the people in the vehicle.

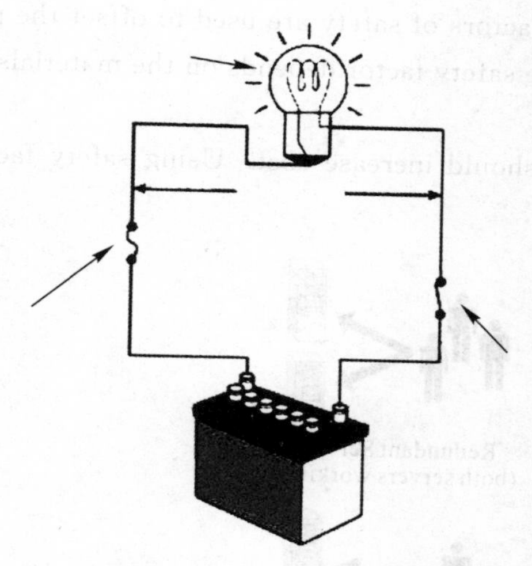

Fig. 1 - 76　Electrical circuits are protected by fuses

Fig. 1 - 77　Crumple zones

Applying the weakest-link, the most important step is to ensure that the weakest link will only fail.

4. Redundancy

Redundancy is the duplication of critical components or functions of a system with the intention of increasing reliability of the system. It can use more elements than necessary to maintain the performance of a system in the event of failure of some elements.

When elements within a system fail, redundancy can avoid that the system as a whole fails. Redundancy is the surest method of preventing system failure. Redundancy can use multiple elements of same or different types. For example, the same information can be presented using of text, audio, and video, a rope is composed of multiple independent strands.

There are two modes of redundancy: active and passive.

(1) Active redundancy

Active redundancy is the application of redundant elements at all times, and can prevent both system and element failure. For example, using multiple independent servers support the performance of system in Fig. 1 - 78. Loads are distributed across all elements such that the load on the each element is reduced.

(2) Passive redundancy

Passive redundancy is the application of redundant elements only when an active element fails, will result in system failure. For example, a spare tire on a vehicle begins to use until the event of a flat tire occur. Passive redundancy is the simplest and most common kind of redundancy.

5. Factor of safety

Factor of safety is the structural capacity of a system beyond the expected loads or actual loads, and the use of more elements than is thought to be necessary to prevent system failure.

Design need to deal with uncertainties. Factors of safety are used to offset the potential effects of these uncertainties. The value of the safety factor depends on the materials and use of the item.

Increasing the safety factor in a design should increase cost. Using safety factors can minimize the probability of failure in a design.

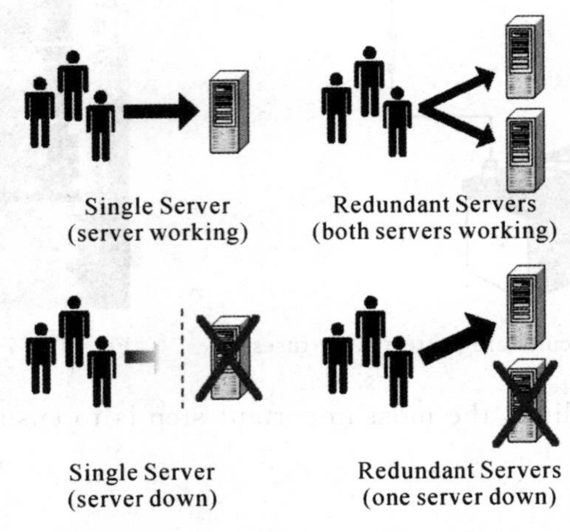


Fig. 1-78 Redundancy design of servers

6. Convergence

Convergence is the tendency for different technological systems to evolve toward performing similar tasks.

Natural or human-made systems that best approximate optimal strategies afforded by the environment tend to be successful, while systems exhibiting lesser approximations tend to become extinct. Convergence is a process in which similar characteristics evolve independently in multiple systems (see Fig. 1-79).

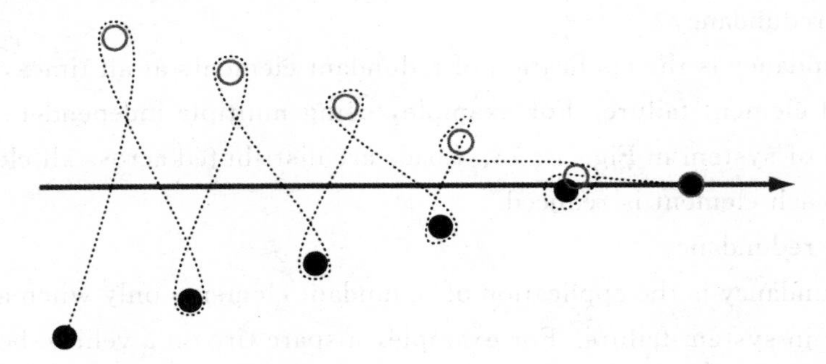

Fig. 1-79 Process of convergence

The degree of convergence indicates its stability so that it is necessary to consider the level of stability and convergence in an environment prior to design. It is very important to focus on variations of convergent designs in stable environments, and explore analogies with other environments and systems for guidance when designing for new or unstable environments.

7. Comparison

Comparison is an examination of two or more items to establish similarities and dissimilarities. People can illustrate relationships and patterns in system behaviors to represent two or more system variables in a controlled way.

There are two kinds of comparisons: apples to apples and benchmarks.

(1) Apples to apples

Comparison data should be presented using common measures and common example (see Fig. 1－80).

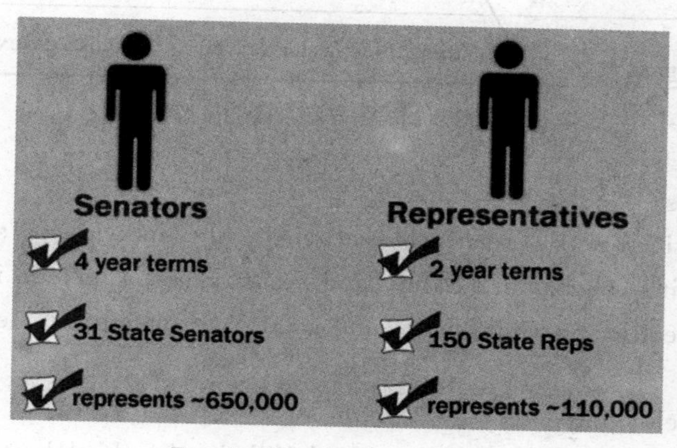


Fig. 1－80　Apples to apples

(2) Benchmarks

Benchmarks can provide a point of reference from which to evaluate the data.

8. Hierarchy of needs

Hierarchy of needs is a theory in psychology, proposed by Abraham Maslow in his paper *A Theory of Human Motivation* in 1943 (see Fig. 1－81). Maslow's hierarchy of needs is often portrayed in the shape of a pyramid, with the largest and most fundamental levels of needs at the bottom, and the need for self-actualization at the top.

(1)Physiological needs

For the most part, physiological needs are obvious. They are the literal requirements for human survival. If these requirements are not met, the human body simply cannot continue to function.

Air, water, and food are metabolic requirements for survival in all animals, including humans. Clothing and shelter provide necessary protection from the elements. The intensity of the human sexual instinct is shaped more by sexual competition than maintaining a birth rate adequate to survival of the species.

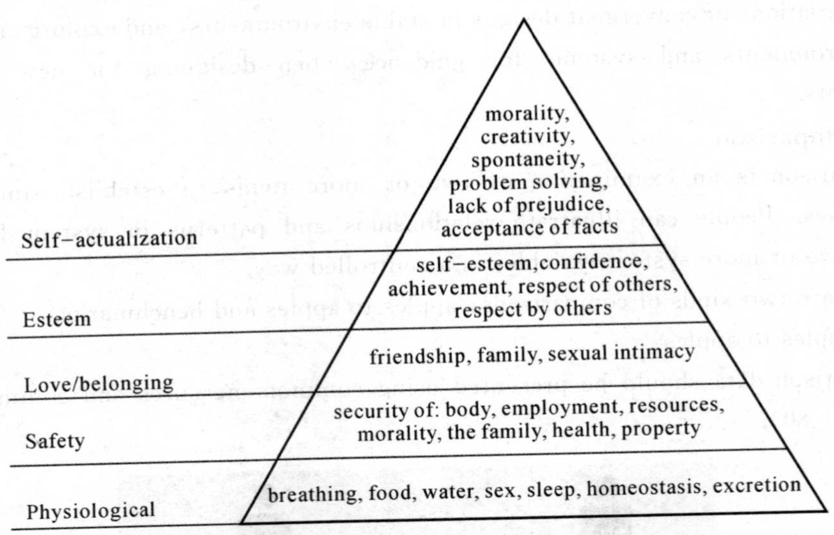


Fig. 1-81 Maslow's hierarchy of needs

(2) Safety needs

After the physical needs relatively satisfied, the individual's safety needs take precedence and dominate behavior. Safety and security needs include in personal security, financial security, health and well-being, and safety net against accidents/illness and their adverse impacts.

(3) Love and belonging

After physiological and safety needs are fulfilled, the third layer of human needs is social and involved feelings of belongingness. Humans need to feel a sense of belonging and acceptance, need to love and be loved by others. In the absence of these elements, many people become susceptible to loneliness, social anxiety, and clinical depression. This need for belonging can often overcome the physiological and security needs.

(4) Esteem

All humans have a need to be respected and to have self-esteem and self-respect. Esteem presents the normal human desire to be accepted and valued by others. People need to engage themselves to gain recognition and have an activity or activities that give the person a sense of contribution, to feel self-valued. Most people have a need for a stable self-respect and self-esteem.

(5)Self-actualization

"What a man can be, he must be." This forms the basis of the perceived need for self-actualization. This level of need pertains to what a person's full potential is and realizing that potential.

In 1970's, hierarchy of needs model increases two layers, including cognitive and aesthetic needs. Cognitive needs include in knowledge, meaning, etc. Aesthetic needs include in appreciation and search for beauty, balance, form, etc.

In order for a design to be successful, it must meet people's basic needs before it can attempt to satisfy higher-level needs.

9. Life cycle

Life cycle identify a set of common stages in the life of commercial products, for example, introduction, growth, maturity and decline (see Fig. 1-82).

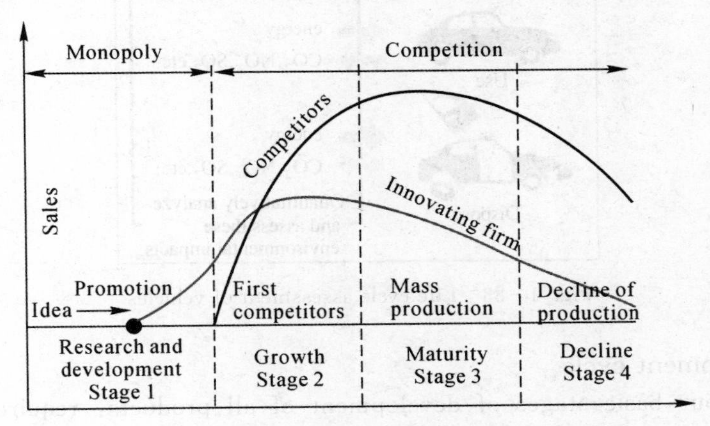


Fig. 1-82 The life cycle

(1) Introduction

The introduction stage is the birth of a product. It will be later stage of the development cycle. The design focus in this stage is to ensure proper performance.

(2) Growth

The design focus in this stage is to scale the supply and demand.

(3) Maturity

The maturity stage is the peak of the product life cycle. The design focus at this stage is to refine the product to maximize customer satisfaction. The next-generation product should begin to develop at this stage.

(4) Decline

The decline stage is the end of the life cycle. Product sales continue to decline and core market share is at risk. The design focus in this stage is to develop transition strategies to migrate customers to new products.

Life Cycle Assessment (LCA) is a tool to assess the potential environmental impacts of product systems or services at all stages in their life cycle (see Fig. 1-83). LCA can be applied in e. g. strategic development, product development and marketing. MITSUBISHI MOTORS introduces the Design for Environment (DfE) concept. This is new process of developing and designing vehicles, incorporates the earth's susceptibility to vehicles through every stage of their life cycle. Vehicles have less environmental impact at every phase. In order to develop vehicles using these guidelines, we will manufacture environmentally friendly vehicles employing LCA methods via our newly introduced quality control system known as "Quality Gate".

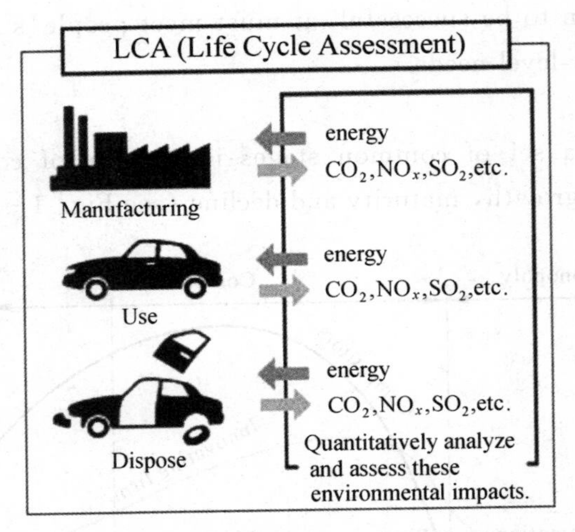


Fig. 1 - 83 Life cycle assessment of vehicles

10. Development cycle

There are four basic stages of development of all products: requirements, design, development, and testing.

(1) Requirements

Requirements are gathered through market research, customer feedback, focus groups, and usability testing.

(2) Design

This stage is design requirements are translated into aspecific form that yields a set of specifications. The goal is to meet the design requirements.

(3) Development

The development stage is design specifications are transformed into an actual product.

(4) Testing

The testing stage is the product is tested to ensure that it meets design requirements and specifications, and will be accepted by the target audience.

# 第2章 设计理念与策略的实证研究 The Case Study of Design Idea and Strategy

Industrial design is the creative activities about the factors concerned with design objects. It is about either tangible object or intangible service and experience. They are not only the external features but also are principally those structural and functional relationships which convert a system to a coherent unity both from the point of view of the producer and the user.

In this part, we will learn from foreign design institutes and companies, and understand how they get design to be applied in a wider scope. By doing so, they will further understand that design means much more than drawing beautiful drafts on paper and making 3D pictures in a computer. For an enterprise, the popularity of its products depends on not only the efforts of its sales and technical staffs, but also the contribution of the designers.

We will see many excellent designing works in this part. While we appreciate these works, we should not stop at their beautiful lines or avant-garde concepts. Behind the exquisite surface of these designs lies the concern for humanity and for the life of the products. They should pay attention to the underlying sophistication and competitive designing concepts and strategies.

Therefore, the highlight of this part is the study of the concepts and strategies of the excellent designs and companies, which is especially important for seniors, who have mastered the basic design methods.

## 2.1 设计组织 Design associations

The role of design associations is to strengthen the connections among designers, promote the development of the design industry, and facilitate the cooperation and exchange between this industry and other industries so as to enable design to better serve the society. Besides, design associations are also obligated to help the public to gain more knowledge of design so that design can get wider recognition.

The purpose of introducing the world's excellent design associations is to inform the readers of the activities and thoughts of foreign designers. Starting from scratch, the industrial design industry of China has developed rapidly; however, limited by the macro environment, it still has many inadequacies in concept and operation. Therefore, it should learn from foreign design associations and ensure that the design community can cooperate with businesses more harmoniously.

The following materials' originals are on the websites of the associations which are listed in the reference at the end of the book.

### 2.1.1 国际工业设计联合会 ICSID （International Council of Societies of Industrial Design）

The International Council of Societies of Industrial Design（ICSID）is a global not-for-profit organization that promotes better design around the world. It was officially founded in 1957. Today，ICSID counts over 150 members in more than 50 countries. ICSID members are professional associations，promotional societies，educational institutions，government bodies，corporations and institutions，which aim to contribute to the development of industrial design.

ICSID facilitates cooperation and interaction among these societies and supports a global network through which design institutions worldwide can stay in touch，share interests，experiences，and resources，as well as provides an international platform for its members to be heard as a powerful voice.

The definition of industrial design was firstly given by ICSID in 1959. Then it changed three times. The first definition was read as follows：

An industrial designer is one who is qualified by training，technical knowledge，experience and visual sensibility to determine the materials，mechanisms，shape，color，surface finishes and decoration of objects which are reproduced in quantity by industrial processes. The industrial designer may，at different times，be concerned with all or only some of these aspects of an industrially produced object.

But the sphere and aim of design are always changing with the development of technology and society. The latest definition of design from ICSID now is as follows：

Design is a creative activity whose aim is to establish the multi-faceted qualities of objects，processes，services and their systems in whole life cycles. Therefore，design is the central factor of innovative humanization of technologies and the crucial factor of cultural and economic exchange.

It should be specially emphasized that in the new definition，design concerns products，services and systems conceived with tools，organizations and logic introduced by industrialization — not just when produced by serial processes. Therefore，the term designer refers to an individual who practices an intellectual profession，and not simply a trade or a service for enterprises.

The four definitions of industrial design by ICSID successfully demonstrate the development stages of industrial design. Given by the first definition，the industrial design is about the appearance design for bulk — production items，which is a narrowly-defined design according to the latest definition. The latest definition of industrial design reflects its various application fields，far-reaching research topics and broad range of influence.

Industrial design not only aims at structure-building，function-undertaking and visual effect displaying，but also tries to provoke thoughts from the aspects of individual，society， culture and environment. For instance, it concerns about the green design on sustainable development and environmental issues, the affiliation of user group and specific user, the relation between production and market, and the cultural diversity under globalization situation.

### 2.1.2　美国工业设计师协会 IDSA (Industrial Designers Society of America)

The Industrial Designers Society of America (IDSA) is the world's oldest and largest design society. The fields include product design, industrial design, interaction design, human factors, ergonomics, design research, design management, universal design and related design fields. The International Design Excellence Award (IDEA), which is called the three most famous international design award with the iF design award and red dot design award, is organized by IDSA annually. The widely published design magazine Innovation is the quarterly by IDSA.

### 2.1.3　香港工业设计师协会　IDSHK (Industrial Designers Society of Hong Kong)

Industrial Designers Society of Hong Kong was founded on 12 June 2002. IDSHK is a regional, non-governmental, non-profit making professional organization formed by a group of enthusiastic competent industrial designers and design educators.

The logo of the society represents the culture of Hong Kong design (see Fig. 2-1). The Chinese Character "品" is the basic element of the logo. It has multiple meaning in Chinese: quality, taste, brand. It symbolizes the uniqueness, professionalism and boundaryless concept of Hong Kong culture. Through modern visual treatments, "品" and "ID" are interlocked together onto a center square which represents China. The opening at left conveys the message of merging to the word with a breakthrough westwards. The outer circle represents the global vision while the layout (square in circle) echoes with the traditional Chinese concept of the universe. Making use of the contrasting colors of orange and grey thus to create the bouncing motions to the eyes.

Fig. 2-1

Alan Yip is an international renown product designer. He found ALAN Yip Design Ltd. in 1990. Throughout these years, over 1,000 products, which cover a wide range of categories have been designed and launched, quite a number of them had hit the million-piece sales record. He also owns over 40 international design and invention patents.

The Sushi series products are his most outstanding design. It is inspired by the Egyptians papyrus and the Chinese bamboo scroll, both of which can be rolled and easy to carry. The rolled calculator is called Sushi calculator (see Fig. 2-2) because it looks like the tools for Sushi making. It is the first calculator in the world, which can be rolled up in such a way. The beautiful color is attractive as modern office supplies. The Sushi calculator achieves many design awards. Derived from the Sushi calculator, the Sushi EL clock (see Fig. 2-3) and Sushi time bracelet (see Fig. 2-4) are also can be rolled. The Sushi series is a good example of how design becomes business.

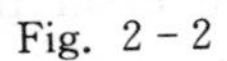

Fig. 2-2

Fig. 2-3

Fig. 2-4

Another famous work of Alan Yip is the ceramic FM radio(see Fig. 2-5). Ceramic is one of the greatest ancient Chinese inventions. The ceramic FM radio is a great combination of ancient ceramic with modern technologies. The shape is distilled from the ancient cooking vessel and embodies both the elegance of ancient vessel and the simpleness of modern design.

Fig. 2-5

Speaking of embedding the Chinese traditional culture elements into product design, there is a lot we can learn from the work of Hong Kong designers. Instead of copying a certain pattern or model from traditional arts, the modern Chinese-style design is inclined to presenting the beauty, spirit and realm of Chinese traditional culture. The combination of abstract cultural meaning and modern design is exactly what the modern Chinese-style design needs.

### 2.1.4 日本工业设计促进会 JIDPO (Japan Industrial Design Promotion Organization)

The Japan Industrial Design Promotion Organization (JIDPO) was founded in 1969. Like other design societies, JIDPO's mission is to promote design by activities among government, industrial bodies, and individual designers. The "Good Design Awards" and "Design Year" campaign are two important activities organized by JIDPO.

The Good Design Awards is a comprehensive program for the evaluation and encouragement of design organized by JIDPO. The most distinguished designs are selected from those submitted for consideration for the Good Design Awards. It is not a beauty contest, nor is it an award that assesses the design's outcome in economic terms. Rather, the Good Design Awards is a system that aims to channel the eminent powers of distinctive designs to build prosperous lives and encourage sound industrial development. Indeed, it is a campaign to brighten and enrich society through design.

Fig. 2-6

Good Design Award winning items can promote its results by attaching "G Mark"(see Fig. 2-6). "G Mark" has been familiar to many people as a trustworthy symbol which connects industry and everyday life.

Japan's industrial design has got an early start and great progress among Asian countries. Besides the support from Japanese government, Japanese industrial design association's own efforts contribute more to the industry development. Campaigns held by industrial design institutes, setting of G-Mark and other incentive measures have strongly stimulated the blossom of Japan's industrial design. G-Mark rewards many different types of design and study. It focuses on the current widely-used design as well as the study of the future prospects closely related to human and social development. By giving rewards, it commends the technology and strength of those great enterprises, at the same time, encourages small and medium-sized businesses and students to make progress.

### 2.1.5 德国设计委员会 GDC (German Design Council)

The German Design Council (Rat für Formgebung) was founded in 1953 to meet the growing need of the business world for information about design. The council provides services for companies, designers, students, media and other involved institutes. Today, the German Design Council is one of the world's leading competence centers for communication and know-how transfer in the design field. With competitions, exhibitions, conferences, consulting, research and publications, they offer new perspectives for representatives of business and design disciplines.

This chapter introduces information of some foreign industrial design institutes, including their establishment, target, mission and major activities. Industrial design organization plays a crucial role in the development of a country's industrial design. On one hand, the industrial design organization guarantees the share of resources and exchange of information by integrating designing resources. Joining the industrial design institute, an individual or division can take precedence to get the industrial design information, participate in industrial design training and communication and get more opportunities for promotion. On the other hand, the organization which represents industrial design has build up a bridge between government, other profession and industrial design, as a result, the public, government and other profession will know and value industrial design. Because their organization actively promotes and popularizes the design knowledge, Japanese people have a deeper understand about industrial design. China' industrial design develops better in the south than what in the north. It is due to not only the human resources and ideology but also the support of government.

## 2.2 设计竞赛 Design competitions

Over all these years, despite the numerous design matches, the 3 major international design contests, namely, iF, Red dot and IDEA, have remained the most influential and convincing. By watching these design contests, they may learn about the cutting-edge design ideas and information, what themes now appeal to the excellent designers of the world, what problems they have engaged to solve, and what they intend their designs to express. Our design concept can thus be renewed and upgraded. Only with a broad vision can one set for himself a greater goal and then reach farther and be more successful on the road of design.

### 2.2.1 国际设计优秀奖 IDEA

The International Design Excellence Awards (IDEA) (see Fig. 2 - 7) is called "the Oscars of Design." It has recognized design excellence in products, sustainability, interaction design, packaging, strategy, research and concepts. The winners include designers and design teams from different countries. They are widely recognized as leaders in the field.

Fig. 2 - 7

The IDEA competition was first held in 1980. In 1991, Business Week magazine began publishing the IDEA winners through its sponsorship of the competition, and it has continued its in-depth reporting every year since. The IDEA articles have produced more than 5 cover stories, including the 2004 report in the European and Asian editions.

### 2.2.2 iF 设计大赛 iF international forum design

iF is a reliable symbolic of good design now. The aim is to strengthen public awareness of design. With the six iF competitions, the iF product design award, the iF communication design award, the iF material award, the iF design award china, the iF packaging award and the iF concept award for students, they have become one of largest and most respected design centers in the world.

The iF logo, which is given to the winners of our competitions, has become an internationally acknowledged symbol of outstanding design. But iF is not only a design competition, but also an independent organization that play an important role in the cooperation of design and economy.

### 2.2.3 红点设计大赛 Red dot design award

The international design competition, the "red dot design award", is aimed at all those who would like to distinguish their business activities through design. The distinction is based on the principle of selection and presentation. Excellent design is selected by competent expert juries in the areas of product design, communication design, and design concepts.

Red dot have three competitions: the red dot product design award, the red dot communication design award, the red dot concept award. The product design competition has existed since 1955. Its award, the "red dot", is an internationally recognized quality seal. The best products receive the "red dot: best of the best" award. The highest award in the communication design competition is the “red dot: grand prix”. The “red dot: junior prize” will be awarded to the best student work.

### 2.2.4　作品分析　Exhibition of the three competitions

(1) Childbirth assistance outside hospitals (see Fig. 2 - 8) ( IDEA 2009, Francisco Lindoro of Umeå Institute of Design )

This concept was designed to aid women and midwives with childbirth in developing countries, where equipment and sterilization are often lacking. Upon opening the sealed container, the user can access sterilized supplies in the sequence of their use: a mat, a lab coat for the midwife, cloths, and a baby sling.

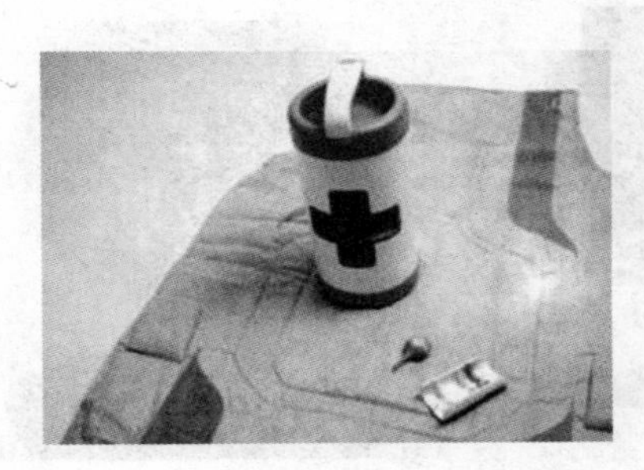

Fig. 2 - 8

Fig. 2 - 9

(2) Energy seed (see Fig. 2 - 9) (IDEA 2009, Sungwoo Park of Samsung Design)

Energy Seed is designed to encourage people to discard batteries in a responsible manner. It is a collection bin for batteries that uses leftover power to light the attached LED streetlamps. When people deposit their batteries, it is akin to planting a seed that later, when it's dark, turns into a flower. It also encourages people to discard their batteries in a responsible manner.

(3) Haptic reader (see Fig. 2 - 10) (IDEA 2009)

The Haptic Reader helps blind people to read non-Braille books. When you place it on the page of a book it scans the letters on the page. The letters are converted to Braille characters, which appear on the surface of the device. The text can also be converted to voice, which is played through the embedded speaker.

Fig. 2 - 10

Fig. 2 - 11

(4) Maptor (see Fig. 2-11) (IDEA 2009)

Unfolding and refolding a paper map is annoying and inconvenient. It is also difficult to find exactly where you are on a paper map. By downloading a map on a Maptor device, you can project a map onto any surface (the wall, ground or hand). The embedded GPS function indicates your exact location.

(5) Li-Ning — Rediscovering sports in modern China (see Fig. 2-12) (IDEA 2011, Ziba Design and Li-Ning)

This is an integrate consumer experience . The design explores from clothing to retail stores to brand identity. It offers Li-Ning an opportunity to develop the youth market. The design team worked on Li-Ning's product strategy, brand identity and retail design.

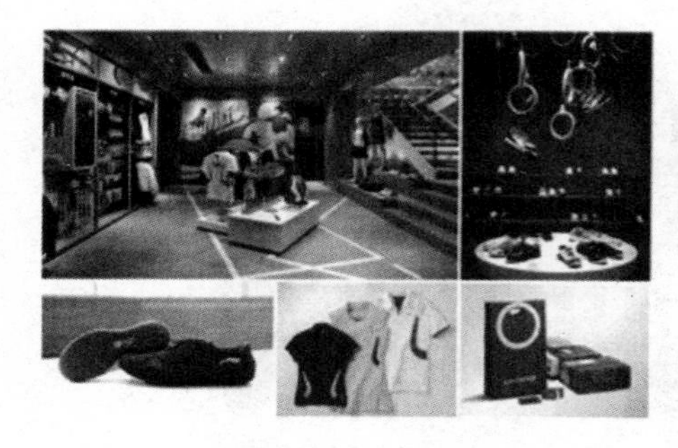

Fig. 2-12

Fig. 2-13

(6) Poetree (see Fig. 2-13) (IDEA 2011, Margaux Ruyant of DSK ISD International School of Design)

Poetree is a funeral urn that evolves over time as a companion through the stages of mourning. The ashes are placed in the urn and are covered with soil. The family takes the urn home and plants a tree in it. When the tree is big enough, it is time to plant the urn in a garden or a park. Eventually the urn, which is made of a biodegradable material and ceramic, will disappear, and only the tree and the ceramic top will remain, just like a gravestone.

(7) In & Out door (see Fig. 2-14) (IDEA 2009)

The In & Out Door helps people determine whether to push or pull the door by directing them to flatten either a protrusion or a hollow on the door. A push panel and pull handle are combined into a set and move organically at the same time. A spring is attached to push panel that returns to its original state.

(8) Medical shower (see Fig. 2-15) (IDEA 2009)

Designed with a compartment to hold traditional Chinese medicine, Medical Shower enables water to be infused with nourishing herbal remedies as it flows out. The medicinal parcel is inserted into the plastic cover on the front of the showerhead, making a medicinal shower as easy as making a cup of tea.

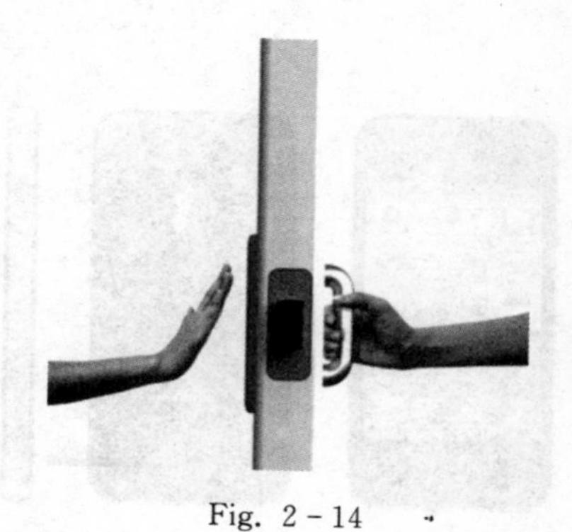

Fig. 2－14

Fig. 2－15

(9) Light breath (see Fig. 2－16) (IDEA 2011)

In critical situations in public, such as in subways, when dense dust or smoke can make it difficult to breathe or see, people often panic, not knowing how to escape. The design of Light Breath is similar to a snorkel respirator to help people recognize how to use it intuitively. They simply hold it between their teeth to breathe in clean air. The attached LED on the front is turned on by the pressure of biting the device, giving people use of both of their hands.

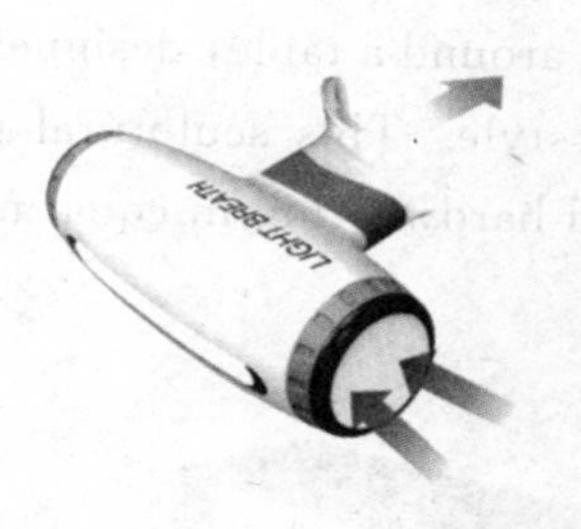


Fig. 2－16

Fig. 2－17

(10) Safe Stop (see Fig. 2－17) (IDEA 2011)

Safe Stop was designed to help people with visual impairments and mobility difficulties to use the bus. Users type the bus number into the kiosk using the Braille keypad; the bus number is then displayed on the kiosk alerting the driver of the passenger's intent to board. The voice notification system also tells the waiting passenger when the bus has arrived.

(11) Tent water collector (see Fig. 2－18) (IDEA 2011, Haimo Bao, Zelong Wang, Yunfei Zhao, Huan Liu, Kui Zhang, Feng Zeng, Yu Fu, Song Qiao and Kun Xu of School of Design, Dalian Nationalities University)

The modular Tent Water Collector provides an easy and effective method to turn rainwater into drinking water. The rain is collected inside the roof, which recesses as the volume of water increases. The water is then funneled through the filtering equipment attached to the underside of the roof and piped directly inside for cooking, drinking and washing.

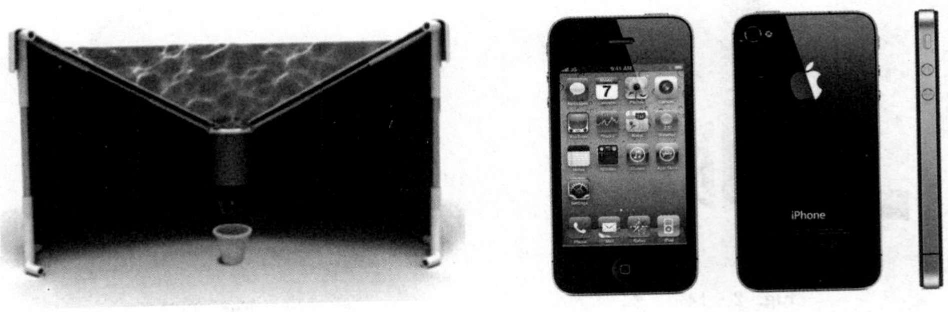


Fig. 2 - 18

Fig. 2 - 19

(12) iPhone 4 (see Fig. 2 - 19) (iF 2009)

The development of the iPhone completely redefined the possibilities of the telephone. It shows us a smart future. Its thin appearance and aluminosilicate material is the outstanding achievement of technics. The innovative multi-touch display offers the user a fascinating surface for cell phones. To ensure the success of a future-oriented product, Apple uses a classical design principle: outstandingly simple, simply outstanding.

(13) Hopper outdoor furniture (see Fig. 2 - 20) (red dot 2011)

Hopper garden furniture is an example of a very successful re-design of the classic beer-garden bench. Based on the traditional "get-together" around a table, designer Dirk Wynants has produced an outdoor product with its very own style. This sculptural piece of garden furniture invites one to sit down and is functional and hardwearing in equal measure.

Fig. 2 - 20

(14) Versatile axis ceramic wall tile (see Fig. 2 - 21) (red dot 2011)

The Versatile collection takes traditional tile design into a new, third dimension. The tile is no longer just a flat surface which covers or protects a wall, but is a spatial form in itself which gives rooms a new perspective. The motif is at once simple and carefully thought through and helps to give an imposing impression of a room.

(15) Chopula cooking spatula (see Fig. 2 - 22) (red dot 2011)

The shape of the Chopula is unusual in every aspect. Visually, this spatula is reminiscent of a swan's foot. The flexibility of the materials used and the shape of the product form a perfect harmony: When it is lifting, turning or cutting food, the spatula adapts to the shape of the pan. With the Chopula, Australian manufacturer Dreamfarm has made the most out of a simple product.

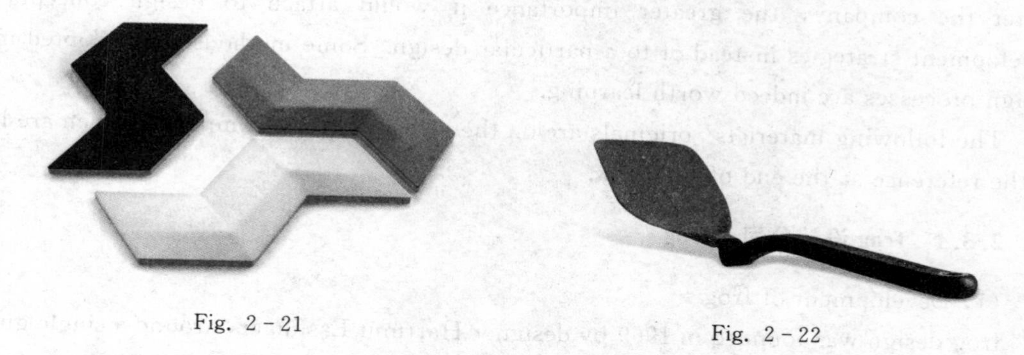

Fig. 2-21　　　　Fig. 2-22

IDEA, iF and red dot have been recognized as the three greatest international design competition. As they have powerful influence throughout the world, the prize-winning works in the competition are proved to be the most excellent. For those designers who take parts in the competition, the award is a certification for ability and also a milestone in career, and for those enterprises which take parts in the competition, the award-winning products are more valuable and competitive in market, hence, the numbers of designers and groups participating in the three competitions have been growing for several years.

What is good design? For a long time, many designers have made various answers to this topic. Dieter Rams summarized ten laws of good design based on his design work experience in BRAUN. IDEO successfully gives an explanation that human-oriented design can be called the good design. The distinctive design philosophy of Apple Inc. allows us to see the commercial value of industrial design. There is no formula or discipline which could leads to good design. Good design is a fruit of hard and long-term study on people, environment and society. Good design, like Haptic Reader which can be read conveniently by both weak sight and blind people, is the satisfaction to unique user groups. Good design, such as BMW 5 Series Touring Passenger Car and Apple product, is the perfect union of craft and art. Good design, like Hopper Outdoor Furniture and Versatile Axis Ceramic Wall Tile, is to show the beauty and quality of materials through the application of new materials in traditional products. Good design, such as Maptor and Chopula Cooking Spatula, is to make our life better, convenient and full of fun by redesigning those articles for daily use and entertainment.

## 2.3　设计企业实例研究　Case study of design companies

China has many design companies, but few of them are doing well. Perhaps they should learn from the development strategies of foreign design companies, especially those outstanding companies with global influence.

This chapter will introduce some prestigious design companies, including their development history and some of their excellent works. The purpose is to stress that the better the company, the greater importance it would attach to design concepts and development strategies instead of to a particular design. Some methods they adopted in the design processes are indeed worth learning.

The following materials' originals are on the websites of the companies which are listed in the reference at the end of the book.

### 2.3.1 frog 设计公司 frog

(1) Development of frog

frog design was founded in 1969 by designer Hartmut Esslinger, around a single guiding principle: "form follows emotion", an adaptation of the familiar phrase "form follows function," declaring that a product's effect upon its user was as vital as its functionality.

In 1981, Steve Jobs began searching for the elusive magic that would give Apple a market edge and he found it in Esslinger's team. A few years later, the Apple IIC(see Fig. 2 - 23) was launched to great fanfare, named "Design of the Year" by TIME Magazine. Apple's revenue soared from $700 million in 1982 to $4 billion in 1986.

1980s, frog cooperated with some companies to established their user experience by product design, engineering, graphic design and package. Its redesign of the Logitech products and brand identity led the computing giant to raise its revenue from $43 million in 1988 to over $200 million in 1995, securing a number-one market position.

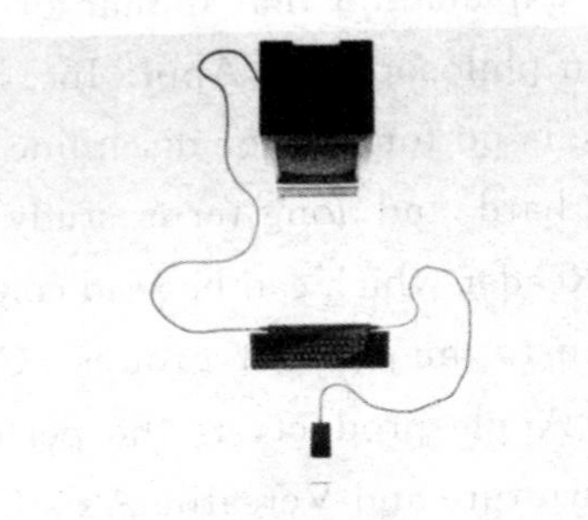

Fig. 2 - 23

Fig. 2 - 24

1990s, after setting up digital media group, frog design has started its professional design in the area of designing webpage, software and user interface for mobile facility. In 2000, the website designed for Dell Computer (see Fig. 2 - 24) by frog design became a role model to e-commerce. In 2001, working with Microsoft, frog design designed interface for Windows XP which was the newest operation system then. It is this interface that has accompanied millions of Microsoft users in life and work. In recent years, frog design has been widely involved in the area of webpage, software, mobile device and consumer electronics to improve user experience by design in the prevalence of computer technology products.

Recently, frog design has expanded the range of services to provide design consultation and long-term development planning for high-end corporate clients. Today, it is no more a company that could only carry out product design, but a company which could offer creative design advice and consultation as well. Under the teamwork of designers, strategy consultants and technical person, frog design has been dedicated to providing the best solutions for well-known global enterprises.

frog design summarized its development to this: We developed from the stage of product design to the stage of design for user interface, and then we stepped into an era of comprehensive design service. The development process of frog design can be regarded as the development process of industrial design. As we mentioned in the previous posts, the evolvement of the definition for industrial design by ICSID indicates the development of frog design. By contrast, some design companies or design department in big enterprises are still in first phase of industrial design. This situation makes industrial design less influential in enterprises and society, which results in its small role in our life.

(2) IDEA Award frog owns

1) Project Masiluleke (see Fig. 2-25). Project Masiluleke means "lend a helping hand" in Zulu. It is a historic partnership between frog design, Pop! tech, iTeach, Praekelt, Aricent, Nokia Siemens and a number of other collaborators. Project Masiluleke is using mobile technology to tackle the worst HIV epidemic in the world in Kwa Zulu Natal, South Africa, where infection rates are over 40%. It combines existing, low-cost diagnostic technologies such as saliva and blood tests with mobile support services in a region where mobile adoption rates are approaching 90%.

frog provides the solution that uses mobile technology in three crucial ways: ① to encourage usage of low-cost diagnostic tools; ② to walk patients through the testing process; ③ to guide them into care should they need it, and encourage healthy preventative behaviors even if they don't.

Fig. 2-25

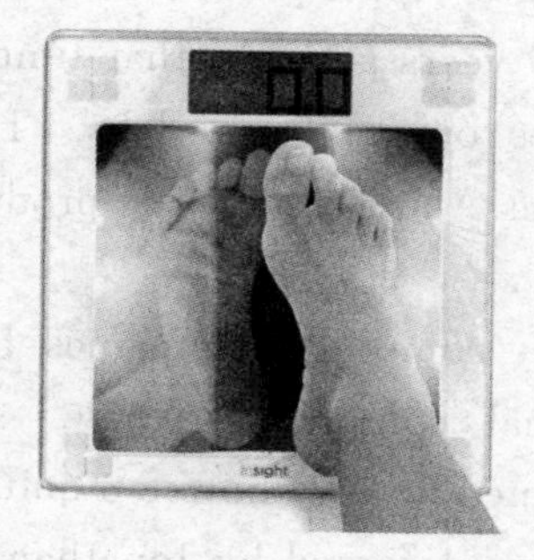

Fig. 2-26

2) Insight Foot Care Scale (see Fig. 2-26). The Insight Foot Care Scale is designed to help patients spot the first signs of foot irregularities and help stop the rapid rise in

amputations. When the consumer steps on the scale, they can see their weight on a large LED readout. At the same time, the system beeps and then uses cue light technology to light up mirrors in the scale, reminding users to check their feet. The scale is designed with angled mirrors and lights that show the consumer a bright and magnified image of the sole of their foot. "The team spent a lot of time on the best angle for the mirrors and light," says John Lapetina, a mechanical engineer in frog's San Francisco studio. "The goal was to provide as clear an image as possible."

3) Intel PoS (see Fig. 2-27)

Intel and frogworked together to combine the retail shopping experience with the convenience of online shopping. The result is an in-store experience for the future that rivals what consumers have grown to love about shopping online: better access to information and features such as reviews or personalized recommendations.

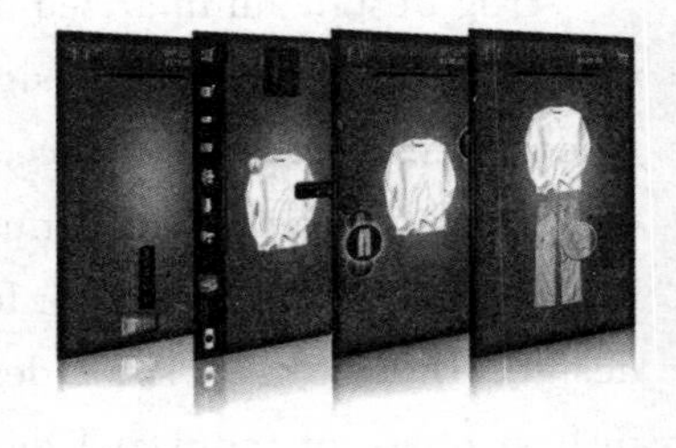

Fig. 2-27

The idea allows shoppers to search inventory directly or with a sales person in real time, and to enter and peruse customer reviews from the store, in essence matching the online shopping experience in a cool kiosk.

(3) The design process

frog has three-part process that helps their business solutions from the sketches to the market. The first process is called discover, which analyze the design research and make it into insight of brands, consumers, and market. The second process is design. Design makes intangible insight becomes tangible products or system. The last process is delivered to the client for implementation.

### 2.3.2 苹果产品设计 Apple

(1) The development of Apple Design

Although now Apple products are known by most people in the world, it was a small workshop thirty years ago. At that time, computers were very big and only used by some science institutes or university labs. The "Big Blue" IBM's main filed were business computer. It was Apple who firstly produced the personal computer and then gave people the concept of "PC".

When most computer companies began to share the cake of personal computer, Apple's iMAC G3 (see Fig. 2-28) came out and showed a quite different visual of computer. It was designed by Jonathan Ive, who designed the later iPhone series. iMAC G3 was distinguished as its shape, color and material. The customers were usually artists, designers and people who had high-income and good

Fig. 2-28

taste.

As everyone knows, the development of Apple owns much to Steve Jobs. In 1980s, Jobs searched for a design partner who could understand his design concept. He found frog design. The two companies developed the "Snow White" design language that was meant to give Apple's products a coherent visual vocabulary, the appearance of being related. The "Snow White " design language were insisted by Apple all along. It was shown in Apple later computer products and iPhone, iPad. This visual vocabulary is featured by very simple appearance, sleek lines and clean colors.

Jobs always declared the importance of industrial design. It was proved by his cooperation with frog design. It was said that there were many times he directly told the industrial designers in his company how the product he wanted looked like and what functions it should have. It is no doubt that Jobs design aesthetics is accepted by most people because now in the market smart phones are all looks like iPhone. So we can say the emphasis on design was there because of Jobs.

(2) Apple's design concepts

When customers see Apple's products, they know it is Apple's design. It is because all Apple's products persists the brand's design concepts. These concepts are visualized as sleek lines, ease of use, environment concern, and the endless innovation.

1) Sleek lines. One of the most exciting design concepts Apple has recently introduced is the unibody. This is the enclosure for the MacBook Air (see Fig. 2 - 29), the 15 inch MacBook Pro, and the new MacBook.

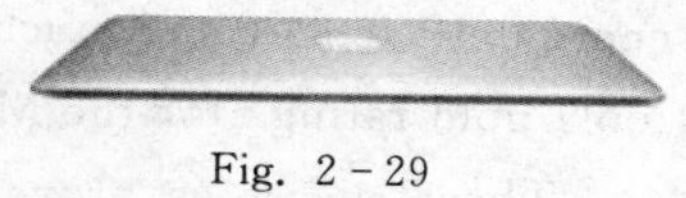
Fig. 2 - 29

Using CNC (computer numerical control) milling machines, a single piece of aluminiumis turned into a laptop casing that accommodates a keyboard, trackpad, ports, display, and the comprehensive interior electronics. It is a simple concept but an extremely difficult idea to achieve in practice. This is why laptop manufacturers generally build their casings from separate parts. These allow room for error. The unibody, however, must be precise in every respect; if not, the interior components won't fit.

This bold approach to design is typical of Apple. Many of Apple's other products have the similar design philosophy. Light weight and slim body are the features of them. It is the result of both industrial design and machine process design.

2) Ease of use. Exterior design alone, however, has not kept Apple's products at the forefront of multimedia technology. The hidden designs and compatibility of the hardware and software within a Mac and iPod have also helped to achieve this.

What behind exterior design are the compatibility of the hardware and the ease of use of the software. This compatibility means that the user of an Apple product has an enjoyable and trouble-free experience. Ease of use extends to the software that comes with every Mac such as the Safari web browser. Safari doesn't just find information quickly on the Internet, which it keeps up with the latest web technology, organizes your data, and helps you distinguish one item from another. Apple's software design specialists never wait for things to happen: they are always searching for ways to develop and improve. Mac users can therefore be confident that their software is up to date.

The iPod touch (see Fig. 2-30), of course, has an outstanding Multi-Touch screen that is a major design feature by itself. Apple also uses the Multi-Touch concept on its iPhone (see Fig. 2-31).

Fig. 2-30

Fig. 2-31

3) Concern for the environment. The third element of Apple's approach to design is concern for the environment. The LED backlighting for displays doesn't contain mercury; internal cables don't have PVC; and components are free of toxic BFRs (brominated flame retardants).

Apple also promotes the recyclable qualities of the aluminium and arsenic-free glass that make up many of the Mac and iPod casings, and has cut back on the packaging it uses.

Such efforts are considerable, and are developing all the time. Among the acknowledgements are the granting of Energy Star status for Macs, thanks to their excellent energy efficiency, and much-coveted EPEAT (electronic products environmental assessment tool) gold ratings for the MacBook Pro and MacBook Air.

These successes prove that environmental concerns can play a major role in design concepts. No doubt other companies will follow Apple's lead.

4) Innovation. Innovation is a permanent principle. With Apple, innovation is fundamental to the company's existence and strategy. Some manufacturers view innovation as an end in itself, but Apple treats it as part and parcel of everyday design and development.

Customers can therefore take innovation for granted. It underwrites products that are great to look at, straightforward to use, and that respect the environment. In other words, they're brilliant designs.

### 2.3.3 IDEO设计咨询公司 IDEO

(1) Introduction

In the early 1990s, David Kelley, Bill Moggridge, and Mike Nuttall, IDEO formed as a new type of collaborative, which soon emerged as a significant leader in the global epicenter of design: IDEO.

The company has many standard-setting successes, such as the first mouse (for Apple

Computer), the first folding notebook computer (for GriD Computer), the first soft-handled toothbrush (for Oral-B).

IDEO has some persisted disciplines such as human factors, business thinking, and organizational design. It provides the service for companies who are seeking to grow and innovate through design. As a result, IDEO's practices have shifted from a methodology of innovation to an underlying culture of innovation to an overarching mental orientation that goes by the name of "design thinking".

Today, IDEO is an award-winning global design firm that takes a human-centered, design-based approach to helping organizations in the public and private sectors innovate and grow.

(2) Design thinking

"Design thinking is a human-centered approach to innovation that draws from the designer's toolkit to integrate the needs of people, the possibilities of technology, and the requirements for business success." — Tim Brown, president and CEO.

Thinking like a designer can transform the way organizations develop products, services, processes, and strategy. This approach, which IDEO calls design thinking, brings together what is desirable from a human point of view with what is technologically feasible and economically viable. It also allows people who aren't trained as designers to use creative tools to solve a vast range of challenges.

Design thinking is a deeply human process that taps into abilities thatthey all have but are overlooked by more conventional problem-solving practices. It relies on our ability to be intuitive, to recognize patterns, to construct ideas that are emotionally meaningful as well as functional, and to express ourselves through means beyond words or symbols. Nobody wants to run an organization on feeling, intuition, and inspiration, but an over-reliance on the rational and the analytical can be just as risky. Design thinking provides an integrated third way.

The design thinking process is best thought of as a system of overlapping spaces rather than a sequence of orderly steps. There are three spaces to keep in mind: inspiration, ideation, and implementation. Inspiration is the problem or opportunity that motivates the search for solutions. Ideation is the process of generating, developing, and testing ideas. Implementation as the path that leads from the project stage into people's lives.

Operating from this perspective, IDEO uses a mix of analytical tools and generative techniques to help clients see how their new or existing operations can look in the future — and build road maps for getting there. Their methods include business model prototyping, data visualization, innovation strategy, organizational design, qualitative and quantitative research, and IP liberation.

All of IDEO's work is done in consideration of the capabilities of the clients and the needs of their customers. As they integrate toward a final solution, they assess and reassess the designs. Their goal is to deliver appropriate, actionable, and tangible strategies. The result: new, innovative avenues for growth that are grounded in business viability and market desirability.

(3) Works

1) Node chair (see Fig. 2 - 32 to Fig. 2 - 35). Node chair is an education solution to improve the classroom experience. The IDEO team observed that tablet-arm desks had remained unchanged for decades, even though class sizes and densities had grown dramatically. This is a requirement of the learning environment that is more free and easily to change.

The final product, dubbed the Node chair, has received praise for promoting student collaboration, allowing educators to reconfigure classrooms to fit different teaching styles, and enabling institutions to save money by making spaces more flexible and accommodating for varied uses.

The Node chair is comfort for students to sit in and adapt their posture. It also swivels, so that students can swing around to look at other students when they are making presentations. The rolling base allows the chair to move quickly to have different seat arrangement. And the desktops can be combined like a conference table when the students are doing group works.

Fig. 2 - 32

Fig. 2 - 33

Fig. 2 - 34

Fig. 2 - 35

2) Project carrot for the centers for disease control (see Fig. 2 - 36). Project Carrot is a program aimed at exploring how environmental changes, health policies, marketing, and communication efforts could be better integrated to deter childhood obesity. It is initiated by CDC (the Centers for Disease Control and Prevention). But the impact of the program was unsatisfied.

Fig. 2 - 36

IDEO helped CDC to reframe the project. The team spoke with teens and adults. They got some insights from the point of teen minds and reframe the project. The final prototypes consisted of an Internet campaign and viral video aimed at teens titled "I Ate What?!"; the Industry Toolkit, a new protocol for public health working with private companies; and a suite of strategic policy tools, including a Health Analysis Impact for policy makers.

This design program is different from the Node chair. The Node chair is a product design to give different experience. Project Carrot is a process that IDEO use its human-centered design methodology to understand the objects. The result is not a real product but some plans and policies.

3) Branch experience for GE money bank (see Fig. 2 - 37). Banking experience is generally characterized by a lack of trust and understanding. Because even customers with highly education have limited knowledge of the services in banks.

Fig. 2 - 37

GE Money Bank see this as an opportunity. They cooperate with IDEO to reinvent its branch concept and in-store experience. The aim is to change the "us vs. them" atmosphere into the desired "we".

The new branch experience reshapes how customers meet with employees, placing the emphasis on service. The spatial layout resembles a restaurant more than a traditional bank; GEMB and IDEO reconceived back-of-house activities and added team rooms and features that make welcoming clients, cash handling, portfolio management, and organization much easier.

4) IDEO toy lab. As a part of IDEO, a human-centered design and innovation firm, Toy Lab believes that design needs to stem from real people, real needs, and real desires.

Because of this, they constantly talk to kids and parents, keeping their eyes open for what's new and important in their lives.

Recently, they noticed a new phenomenon in play: the iPhone. They found it compelling that adults weren't the only ones loving this new toy and decided to explore the iPhone apps for kids space.

While kids are interacting with a number of different iPhone apps every day, surprisingly few of these applications are designed with this precocious demographic in mind. They put together a team consisting of child development experts, veteran toy designers, and interaction designers to uncover the real potential of iPhone apps for kids for the preschool set, ages 2 through 6. Through interviews and observations, they discovered compelling insights that left little doubt that there was fun to be had in this field.

The app "Balloonimals" (see Fig. 2-38) is designed for preschool kids to practice their fingers and creativity by blowing, shaking and touch a balloon and make it into different animals. The pop of the balloon give more surprise to kids.

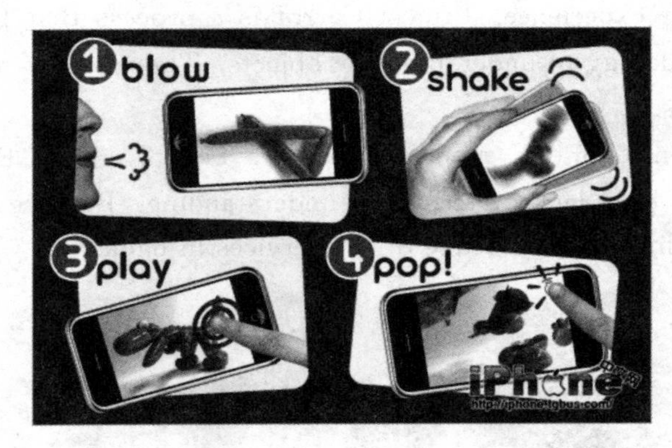


Fig. 2-38

(4) HCD tookit

Human-centered design is a process that has been used for decades to create new solutions to design challenges. The process helps people hear the needs of the people and communities they're designing for, create innovative approaches to meet these needs, and deliver solutions that work in specific cultural and economic contexts. Centered in optimism and embracing constraints and complexity, the HCD process helps users ask the right questions. Ultimately, it can increase the speed and effectiveness of implementing solutions that have an impact on the lives of the people these solutions were designed for.

The Human-Centered Design Toolkit was designed specifically for people, nonprofits, and social enterprises that work with low-income communities throughout the world. The HCD Toolkit walks users through the human-centered design process and supports them in activities such as building observation and empathy skills, prototyping, leading workshops, and implementing ideas. The HCD Connect platform represents the evolution of the HCD Toolkit. People using the HCD Toolkit or human-centered design in the social sector now have a place to share their experiences, ask questions, and connect with others working in similar areas or on similar challenges.

The HCD toolkit includes three phases of the HCD process:

HEAR: Determine who to talk to, how to gather stories, and how to document your observations.

CREATE: Generate opportunities and solutions that are applicable to the whole community.

DELIVER: Take your top solutions, make them better, and move them towards the implementation.

Every phase have tools and tips to help people master the HCD process. They are presented by graphics. Every tip or tool is introduced about the time it cost, the difficulty, the materials it needs, and the participants involved.

For example, one method for the phase of hear is in-context immersion (see Fig. 2-39). This method is to meet people where they live, work and socialize to provide new insights and unexpected opportunities. It usually costs 2 to 4 days. The difficulty is five stars, which means it's the most difficult method. The materials it needs are notepad, pen, camera and small gifts.

Fig. 2-39

By being with people in their real settings and doing the things they normally do, you can talk to them about their experiences in the moment. By immersing yourself in their context, you'll gain empathy and come to understand the people you are designing for on an intellectual and experiential level. This understanding will help you to design solutions with their perspective in mind.

## 2.4 设计思考 Design thinking

In some cases, a product becomes the basis for a new company, while in other situations, new products are developed within an existing company. The former example is "JOSEPH Company". The twin brothers bought a design of a chopping block from MOMA

(Museum of Mordern Art). They redesigned the color and produced it. This product is the basis of their company. The later examples are often seen in modern companies. frog helped Logitech to develop their revenue by design the children's mouse and package. Alan Yip worked for Li Ning to give consultancy about the new brand identity. Jobs saved Apple by iMac G3 in 1998.

The leaders of the contemporary companies in China need to be aware of the relationship between a company's overall brand strategy and how that affects the new products it develops. If a company establish their brand by accident rather than by design, it becomes difficult to compete with other brand.

A good brand often brings good experience to users. A product experience includes both the expression of the product and the interaction with the products. Designers and companies know the importance of user experience now. But before the 1980's, industrial design was more about the external features, structures and functions of the objects produced by industry. Although the point of view of the producer and user were considered, it was not as important as it is now. The methods and theories about human centered design were little.

The development of computer technology changed industrial design to some extent. There are three stages of the computer technology use:

The first stage is the enthusiast stage. In this stage, very few people have the ability of using a computer. In China, most people had never seen a computer in the 1980's. Computers were used only in some labs of universities or research institutes. But the minority users were attracted by the amazing technology. They became the enthusiasts user of this new technology. Enthusiasts don't care if the technology is easy or hard to use because they're so excited by the technology itself or by what it will do for them. They want it, however difficult it is to use.

The second stage is the professional stage. Those who use the technology are often not those who buy it. At this stage, indeed, some people even have a vested interest in the technology being difficult because they're selling their ability to use it; the harder it is, the more valuable their skills are.

The third stage is the consumer stage. People now are less interested in the technology in itself than in what it can do for them. They don't want to spend much time learning how to use it and hate being made to feel stupid. So if it's hard to use, they won't buy it. It's no longer used only by professionals but by a wide range of nonexperts, who just want to use it to pursue their everyday lives. And they're not obliged to use our products: if they can't make them work, they take them back to the store.

We are in the stage of consumer stage. That is why user experience is of great vital to the industrial design. At the core of successful design is the successful experience of using a product or a service. In the former examples, IDEO has its unique method card and toolkit to master the users' needs and improve the user's experience. Apple's products have great influence to people's life. The main reason is its totally new user experience in cell phone and computer. The computer technology is combined with industrial design perfectly, offering the users a new way that used a cell phone they never had.

As the end of this chapter, the following words can describe what the industrial designers should do. It comes from Gillian Crampton Smith, the director of Interaction Design Institute Ivrea.

"*Today we need to design computer technology differently, to make it a graceful part of everyday life, like the other things we own: our clothes, the plates we eat off, the furniture we buy for our houses. We've come to a stage when computer technology needs to be designed as part of everyday culture, so that it's beautiful and intriguing, so that it has emotive as well as functional qualities.*

*The challenges designers face is to make this powerful technology fit easily into people's everyday lives, rather than forcing their lives to fit the dictates of technology.*"

— Gillian Crampton Smith

# 第3章 产品创新设计
# Product Innovation Design

科学技术的最基本特征就是不断进步、不断创新。创新是人类文明进步的原动力。创新对人类科学的发展产生了巨大影响,而科学的发展则成为推动人类社会进步和社会变革的第一动力。

人类历史上有无数的发现、发明和创新对人类的生产、生活产生了非常深远的影响,极大地推动了生产力的发展,人们的生活水平不断地得以提高。一谈到创造发明、发现,人们开始可能会认为是很神秘的事,以为创新发明是学者专家的专利品,一般人很难办到,那么我们先看看下面这些实例吧。

传说鲁班在山上砍柴时,一不小心手被草割破了,一般人可能会自认倒霉,而他却对此产生了好奇心,他仔细观察这种草后,发现这草的边上有一排锯齿,根据这个发现,鲁班发明了至今仍在使用的锯子;传说瓦特在观察到水烧开后蒸汽能将壶盖顶起,依据这个原理他产生了蒸汽动力的设想,并最终发明了蒸汽机,促成了第一次工业革命;还有大家非常熟悉的阿基米德在洗澡时发现浮力定理,从而检验出皇冠是否为纯金的故事;牛顿从树上的苹果会掉下来的现象发现了万有引力。

我国科技在原始性创新方向的不足在近些年越来越突显。在科学技术领域,国家自然科学奖、国家发明奖连续多年空缺;在产业技术领域,我国的发明专利只有美国和日本的1/30,韩国的1/4。

创新势在必行,撒切尔夫人曾说:"现代工业设计要比首相更重要。"杨振宁曾预言:"21世纪将是创新的世纪。"创新不是任何人、任何国家的专利,只要我们提高全民族的创新意识,普及教育各种创新思维和创新技法,必将迎来中华民族灿烂的一页。

## 3.1 创新设计 Innovation design

设计是人类社会最基本的一种生产实践活动,它是创造精神财富和物质文明的重要环节。创新设计是技术创新的重要内容。

工程设计是建立技术系统的第一道工序,它对产品的技术水平和经济效益起决定性的作用。据统计,产品成本的75%~80%是由设计阶段确定的。

设计的本质是革新和创造。强调创新设计是要求在设计中更充分发挥设计者的创造力,利用最新科技成果,在现代设计理论和方法的指导下,设计出更具有竞争力的新颖产品。

### 3.1.1 设计与创新 Design and innovation

1.设计

设计是什么?实际上,设计本身就是一种创造,是人类进行的有计划的活动。

设计的发展与人类历史的发展一样，是逐渐进化、逐步发展的。例如，人类开始的设计是一种单凭直觉的创造活动。这些活动的意义仅仅是为了满足生存：为了保暖就剥下兽皮或树皮，稍加整理就披在身上防寒，也就设计了服装；为了猎取动物，分食兽肉，就设计了刀形斧状的工具，这也许就是最初的结构设计。

后来设计就发展了，不再仅仅是为了生存，而上升到为了生活质量的提高和精神上的某种需要的满足。人们开始利用数学与物理的研究成果解决设计问题。在设计的产品经过实践的检验，并有了丰富的设计经验以后，就开始归纳总结出各种设计的经验公式；还通过试验与测试获得各种设计参数，作为以后设计的依据。同时开始借助于图纸绘制设计产品，逐步使设计规范化。

现在的设计或称现代设计，则不论深度还是广度都发生了巨大的变化。已不再把时间花费在烦琐的计算与推导上，平面图纸的设计也逐渐被取代，出现了优化设计、并行设计和虚拟设计等。设计产品更新换代的时间逐渐缩短，第一代产品刚问世不久，第二代、第三代产品则很快会接踵而来。例如，自1790年美国实施专利制度以来，至今已有600多万件专利。前100万件用了85年；后100万件只用了8年。在最近8年里，平均每天产生专利300件。在这样一个迅猛发展的时代，人们的要求越来越高，也就对设计以及设计工作者提出了更高的水准。设计向什么方向发展，设计如何解决现代人的需求，已经成为重要的话题。

2.创新

“创新”一词一般认为是美国一位经济学家舒彼特最早提出的。他把创新的具体内容概括为五个方面：①生产一种新产品；⑧采用一种新技术；③利用或开拓一种新材料，④开辟一个新市场；⑤采用一种新的组织形式或管理方式。他同时指出：“所谓创新是指一种生产函数的转移。”

概括地说，创新就是创造与创效。它是集科学性、技术性、社会性、经济性于一身，并贯穿于科学技术实践、生产经营实践和社会活动实践的一种横向性实践活动。

创新理论体系主要内容包括技术创新、意识创新、制度创新、市场创新和管理创新。

技术创新为主导地位，作为一个国家、地区或企业，它的存在或竞争实力的大小，经济发展和社会进步的程度，最终取决于技术创新，其他创新活动均为技术创新服务。

意识创新起先导作用，没有创新意识也就没有创新活动。

制度创新起保证和促进作用，即促进技术创新。自改革开放以来，中国的经济体制已逐步由计划经济体制转向社会主义市场经济体制，这为技术创新创造了良好的宏观环境。

市场创新起导向和检验作用。通过市场竞争迫使、激励企业不断创新。市场把创新成功与否的裁决权交给消费者，由消费者的需求引导创新的方向，检验创新成功与否。

管理创新具有协调、整合创新系统各要素的作用。

### 3.1.2　设计过程　Design process

设计一般分为产品规划、方案设计、技术设计、施工设计等四个阶段。

产品规划阶段进行详细的需求调查、市场预测，确定设计参数和制约条件，最后给出详细的设计任务书（或要求表）作为设计、评价、决策的依据。

方案设计（也称概念设计）阶段确定产品的工作原理，并对产品的执行系统、原动系统、传动系统、测控系统等作方案性设计，将有关机械机构、液压线路、放电控线路用简图形式表达。

技术设计阶段在原理方案基础上进行具体结构化设计，选材料，定零件的构形和尺寸，进行各种必要的性能计算，最后画出部件的装配草图。为了提高产品的竞争力还需应用先进的设计理论和方法提高产品的价值（改善性能、降低成本），进行产品的系列设计，考虑人—机工程原理提高产品的宜人性，利用工业美学原则对产品进行更好的外观设计等，使产品既实用，又适应市场商品化的需要。这阶段往往要通过模型试验检验产品的功能原理和性能。

施工设计阶段进行零件设计和部件装配图的细节设计，完成全部生产图纸并编制设计说明书、工艺卡、使用说明书等技术文件。

正式投产前产品的试制将检验加工工艺和装配工艺，并进行较详细的成本核算，从而提出修改意见，进一步完善整个产品的设计。

计算机的应用大大推进了设计速度。目前可通过计算机辅助设计进行方案设计、技术设计并绘制图形。或者直接输出信号，进行产品的数控切削加工或成形加工，但不论采用什么设计手段，合理掌握设计过程，抓住每个设计阶段的特点和重点，有利于调动设计人员的创新精神，提高产品的设计质量。

### 3.1.3 创新设计的类型 Classification of innovation design

根据设计的内容特点，创新设计可分为开发设计、变异设计和反求设计三种类型。

1）开发设计。针对新任务，提出新方案，完成从产品规划、原理方案、技术设计到施工设计的全过程。

2）变异设计。在已有产品的基础上，针对原有缺点或新的工作要求，从工作原理、机构、结构、参数、尺寸等方面进行一定变异，设计新产品以适应市场需要，提高竞争力。如在基本型产品的基础上，开发不同参数、尺寸或不同功能性能的变型系列产品就是变异设计的结果。

3）反求设计。针对已有的先进产品或设计，进行深入分析研究，探索掌握其关键技术，在消化、吸收的基础上，开发出同类型的创新产品。

开发设计以开创、探索创新，变异设计通过变异创新，反求设计在吸取中创新，创新是各种类型设计的共同点，也是设计的生命力所在。为此，设计人员必须发挥创造性思维，掌握基本设计规律和方法，在实践中不断提高创新设计的能力。

### 3.1.4 创新设计的特点 Characteristics of innovation design

创新设计属于技术创新范畴。对创新设计的要求要比对设计的要求提高了许多。创新设计不仅是一种创造性的活动，还是一个具有经济性、时效性的活动。同时创新设计还要受到意识、制度、管理及市场的影响与制约。因此，需要研究创新设计的思想与方法，使设计能继续推动人类社会向更高目标发展与进化。归纳起来创新设计具有如下特点：

1.独创性

创新设计必须具有其独创性和新颖性。

设计者应追求与前人、众人不同的设计方案，打破一般思维的常规惯例，提出新功能、新原理、新机构、新材料，在求异和突破中体现创新。

2.实用性

创新设计必须具有实用性。纸上谈兵无法体现真正的创新。

发明创造成果只是一种潜在的财富。只有将它们转化为现实生产力或市场商品，才能真正为经济发展和社会进步服务。1985—1995 年中国发明协会向社会推荐和宣传的发明创造成果有 11 000 多项，其中只有 15%转化为生产力；而这 10 年中我国的专利实施率仅为25%～30%，看来专利、科研成果和设计的实用化都是需要解决的问题。

设计的实用化主要表现为市场的适应性和可生产性两方面。

市场适应性指创新设计必须针对社会的需要，满足用户对产品的需求。某厂设计一种新型多功能机床，采用了不少新结构。但当时市场此类机床过剩，产品无法推出，设计趋于流产。20 世纪 70 年代，科学家已开始发现氟利昂会破坏高空臭氧层对紫外线的吸收，并影响到人类的生活。上海第一冷冻机厂较早地抓住制冷设备中的这个关键问题，积极设计研制新原理的溴化锂制冷机，代替原来大中型空调机上的氟利昂制冷设备，这种创新设计取得成功，并带来巨大经济效益。

可生产性要求创新设计有较好的加工工艺性和装配工艺性，能以市场可接受的价格加工成产品，并投入使用。

3. 多方案选优

创新设计从多方面、多角度、多层次寻求多种解决问题的途径，在多种方案比较中求新、求异、选优。

以发散性思维探求多种方案，再通过收敛评价取得最佳方案，这是创新设计方案的特点。

如打印设备多年来一直沿用字符打印，虽有各种形式，但很难提高打印速度。随着计算机的发展，推出通过信号控制进行点阵式打印的新模式，引起打印设备领域的一场革命。点阵打印一开始采用针式打印机，完全是机械动作，结构复杂，要经常维修，打印清晰度也不够理想。近年来不断推出不同原理的喷墨式、激光式、热敏式打印机，正是在多种方案的比较中得到了各种符合市场需要的新型打印设备。

## 3.2　创新教育与创新人才
## Innovation education and innovative talent

### 3.2.1　创新教育　Innovation education

知识是创新的前提，没有知识就不可能掌握现代科学技术，也就没有创新能力。因此，教育是提高创新水平的重要手段。

联合国教科文组织的一份报告中说：“人类不断要求教育把所有人类意识的一切创造潜能都解放出来。”即通过教育开发人的创造力，要求和期盼教育在创新人才培养中承担重要任务。联合国教科文组织也做过调研，并预测 21 世纪高教五大特点。

1)教育的指导性——打破注入式和用统一方式塑造学生的局面，强调发挥学生特长，自主学习，教师从传授知识的权威改变为指导学生的顾问。

2)教育的综合性——不满足传授和掌握知识；强调综合运用知识、解决问题能力的培养。

3)教育的社会性——从封闭校园走向社会，由教室走向图书馆、工厂等社会活动领域，开展网络、远程教育。

4)教育的终身性——由于知识迅速交替，由一次性教育转变为全社会终身性教育。

5)教育的创造性——改变教育观，致力于培养学生创新精神，提高创造力。

根据以上特点，中国高等教育人才培养也正开展由专才性向通才性过渡，努力培养并造就出大批具有创新精神与创新能力的复合型创新人才。

如何培养与造就一大批高素质的创造性、开拓型人才则是创新教育必须面对的问题。首先必须更新教育思想和转变教育观念。教育不仅是教，更重要的是育。教也不只是传授传统的知识，还要传授如何获取知识。育就是培育、培养、塑造。其次要探索创新的人才培养模式。不只是在课堂上教，在学校里教，要走出教室，走向社会。积极组织学生开展课外科技活动与社会实践，给学生创造一个良好的探究与创新的条件与氛围。当然还要注重教学内容的改革与更新。在教育中，发明创造的观念、创新的能力是与知识同样重要的内容。进行创新设计的学习不仅是传授一些创新技法，而且要激发学生的兴趣，产生主动获取知识的愿望；同时还要培养善于思维、善于比较、善于分析和善于归纳的习惯。

### 3.2.2 创新人才的特点 Characteristics of innovation talent

创新人才应具备下述特点：

1)有如饥似渴汲取知识的欲望和浓厚的探究兴趣。这样才能发现问题，提出问题，解决问题，并形成新的概念，做出新的判断，产生新的见解。

1930 年诺贝尔医学奖获得者芬森就是一例。丹麦科学家芬森到阳台乘凉，看见家猫却在晒太阳，并随着阳光的移动而不断调整自己的位置。这样热的天，猫为什么晒太阳？一定有问题。带着浓厚的探究兴趣.他来到猫面前观察，发现猫身体上有一处化脓的伤口。他想，难道阳光里有什么东西对猫的伤口有治疗作用？于是他就对阳光进行了深入的研究和试验，终于发现了紫外线——一种具有杀菌作用，肉眼看不见的光线。从此紫外线就被广泛地应用在医疗事业上了。

2)具备强烈的创新意识与动机和坚持创新的热情与兴趣。只有这样深入钻研，紧追不舍，才能确立新的目标，制定新的方案，构思新的计划。

许多科学家正是带着这种强烈的责任感与使命感，做出了重要的贡献。法国的细菌学家卡莫德和介兰，为了战胜结核病，经历了 13 年的艰苦试验，成功地培育了第 230 代被驯服的结核杆菌疫苗——卡介苗。

3)具备创新思维能力和开拓进取的魄力。这样才能高瞻远瞩，求实创新，改革奋进，并开辟新的思路，提出新的理论，建立新的方法。

4)具备百折不挠的韧劲与敢冒风险的勇气和意志。这样才会正视困难和重视困难，并创造出新的道路，迎接新的挑战，获取新的成果。

### 3.2.3 创新人才的培养 Innovation talent training

1.培养创新意识，唤醒、挖掘、启发、解放创造力

1)创造力具有普遍性。相信人人具有创造力。创造学的基本原理告诉人们，创造力是每个正常人都具有的一种自然属性。心理学研究也表明，一切正常人都具有创造力，人人都可以搞发明创造。许多“小人物”搞发明的故事，给了我们很多启示。

诺贝尔物理学奖获得者詹奥吉说：“发明就是和别人看同样的东西却能想出不同的事情。”我国著名教育家陶行知先生在“创造宣言”中提出“处处是创造之地，人人是创造之人”，鼓励人们破除迷信，并敢于走创新之路。

世界充满矛盾，现实不一定合理。善于观察事物、发现矛盾和需要，这往往是创造的动力和起点。如齿轮是重要的机械零件，渐开线齿轮精度测量项目多，要采用多种测量仪器，长期以来一直是加工、使用中的难点。成都某研究所针对此矛盾设计开发了齿轮全误差测量仪，通过被测齿轮与标准齿轮啮合时角速度等参数的变化，直接分析测量齿轮的各项误差，大大简化了测量过程。

创造需要付出非凡的劳动，需要有坚定的毅力，克服重重的困难才能取得胜利。爱迪生研究白炽灯时，为寻找灯丝材料曾用过6 000多种植物纤维，试验1 600多种耐热材料。另外，666农药因试验666次才得到成功，故取此名。

人的创造力通过教育和训练是可以提高的。

创造力以心理活动为主，而心理活动的生理基础和物质基础是大脑和以大脑为核心的神经系统。揭示脑生理机制的奥秘就可以证明人人具有创造力。研究表明，人的智力、创造力取决于人脑神经元的构造，每个神经元之间的"触突"依靠电-化学反应形成了某种联系，思维就在这电-化学反应中进行。一瞬间有10万～100万个化学反应发生。人脑有140亿个神经元，它们之间联系高达$10^{783\,000}$单位，作用远远超过任何超级大规模集成电路。

目前人脑还有极大潜力未开发。神经生理学家认为，一般人的大脑潜力仅利用了4%～5%，少数人利用了10%左右。爱因斯坦的大脑的质量和细胞数量与常人相仿，但神经细胞的"触突"比常人多，说明它的大脑开发得比别人多，但最多也仅达30%。可见人的大脑潜力极大，开发创造力大有可为。

2)培养观察能力。培养善于观察事物、发现问题的能力。

观察能力是对事物及其发展变化进行仔细了解，并把其性质、状态、数量等因素描述出来的一种能力。

具有敏锐的洞察力，善于发现已有的事物或原理，用以解决矛盾，这也是创新意识的体现。山东有位叫王月山的炊事员观察到灶里的煤火燃烧不旺时，只要拿根铁棍一拨，火苗从拨开的洞眼窜出，火一下就旺起来。后来，他用煤粉做煤球、煤饼时，主动在上面均匀地戳几个通孔，这样不仅使火烧得旺，而且节省燃煤。大家熟悉的蜂窝煤就是这样发明的。科学家张开逊在调试某仪器时，发现每当有人进入房间时，仪器零点即发生飘移。他抓住这个现象，经过多次试验和深入研究，探讨出是气流温度场的作用。在此基础上他根据人的呼吸影响气流温度场和气流密度变化的现象，开发出高分辨度测温仪，精度达1/1 000摄氏度，用于新生儿和重病人的呼吸监护，效果很好。

影响观察能力的因素是感觉器官及已有的知识和经验。

提高观察能力的途径是培养浓厚的观察兴趣；培养良好的观察习惯（即观察要有目的性、计划性、重复性，观察结果要做记录等）；培养科学的观察方法（观察时要注意全局与局部、整体与细节、瞬间与持续现象的关系等）；时刻做有心人，时刻让感觉器官处于积极状态。

发现能力是指从外界众多信息源中发现自己所需要的有价值问题的能力。发现问题的能力不仅仅在于发现，而更应注重对所发现问题的各种信息的融合贯通，理清它们的来龙去脉，为解决问题提供重要信息。

历史和实践表明，科学上的突破，技术上的革新，艺术上的创作，无一不是从发现问题、提出问题开始的。爱因斯坦认为，发现问题可能要比解答问题更重要。

3)培养良好的创造心理。创造力受智力与非智力因素影响。

智力因素包括观察力、记忆力、想象力、思考力、表达力、自控力等。

非智力因素包括信念、情感、兴趣、意志、性格等。

前者是基础,后者是动力。例如,兴趣对观察力与注意力具有很大影响,只有对某事物极感兴趣,才会注意它、观察它,也才会从中发现问题并解决问题。情感是想象的翅膀,丰富的情感可以使想象更加活跃,而想象又可以使人的创造精神充分发挥。意志是一种精神力量,它使人精神饱满,不屈不挠,不达目的誓不罢休。教育者应充分运用信念、情感、兴趣、意志、性格等非智力因素,开发与调动被教育者内在的积极因素,使他们通过对非智力因素的培养,促进智力因素的发展与提高。

2.加强创造实践

要设置一系列的实践环节,进行创造活动训练。例如美国通用电气公司对有关科技人员在开设创造课程的同时,还进行了一些创造实践的训练,两年后取得很好的效果,按专利量计,人的创造力提高了5倍。

在学校里,开设创新设计类课程,开设创新设计实验室,开发创新设计的实验,为学生创造一个良好的创新实践环境,对于培养和塑造具有创新能力的学生是极其有效的。另外,大学生的各种课外科技活动竞赛也是很好的创造实践活动,其中不少作品在学科中具有突破性的意义。

3.排除各种影响创新活动的障碍

(1)环境障碍

外部环境有自然环境与社会环境。

社会环境包括文化条件(守旧意识、中庸之道、平均主义等)与社会制度(计划经济下的等、靠、要,僵化的人事制度和应试教育等)。

内部环境包括心理、认知、信息、情感、文化等不利的个人因素。

(2)心理障碍

从众心理是指个人自觉或不自觉地愿意与他人或多数人保持一致的个性特征,是求同思维极度发展的产物,俗称随大流。一般来说,普通人从10岁以后,开始出现从众心理,会有意无意地同周围人尽量保持一致。

法国一位心理学家曾做过一个试验,他让几位合作者扮成在医院候诊室等待看病的病人,并让他们脱掉外衣,只穿内衣裤。当第一个真正的病人来时,先是吃惊地看了这些人,思索一会儿,然后也脱掉自己的外衣顺序坐到长凳上,第二个病人,第三个病人……竟无一例外都重复了同样的行为,表现出惊人的从众性。

从人的心理特征来看这个例子,说明当与别人一致时,感到安全;而不一致时感到恐慌。从众倾向比较强烈的人,认知、判定时,往往符合多数,人云亦云,缺乏自信,缺乏独立思考的能力,缺乏创新观念。

法国一位科学家也做过一个有趣的试验,它把一些毛毛虫放在一个盘子的边缘,让它们头尾相连,一个接一个,沿着盘子边缘排成一圈。于是,这些虫子开始沿盘子爬行,每一只都紧跟着前面的一只,不敢走新路,他们连续爬了七天七夜,终因饥饿而死去。而在那个盘子中央,就摆着毛毛虫爱吃的食物。从这个试验中,可以看出动物也具有这种心理特征。

偏见与保守心理指个性上的片面性与狭隘性,对新事物的反感与反抗。有这种个性特征的人在看待任何事物时,往往是先入为主,在头脑里形成对问题的固定看法,用先前的经验抵制后来的经验；对逐渐出现的变化反应迟钝，不愿意接受新事物；在思维上代表了封闭性与懒惰性。

国外一位心理学家做过一个试验，他先让受试者看一张狗的图片，然后再让受试者看一系列类似狗的图片，其中每一张图片都与前一张有差异，即每一张都减少一点狗的特征，增加一点猫的特征。这些差异累积起来，使最后一张图片像猫而不像狗。偏见与保守的人则一直认为图片里的是狗，而不是猫；而思维灵活的人则早认出图片里的已经变为猫了。

(3)认知障碍

思维定势指习惯固定模式，机械套用，阻碍新点子产生。有些人如饥似渴地学习知识，积累知识，但运用知识时，却难以突破原有知识的框架，不敢越雷半步。美国心理学家贝尔纳认为，构成我们学习的最大的障碍是已知的东西，而不是未知的东西。

功能固着是指受事物的经验和功能的局限，不可能发现其潜在功能。例如茶杯的功能是作为容器盛水用，不能发现其还可用于画圆，用作量具，甚至可当武器使用。

结构僵化指认知上受物体结构局限，不能发现可变化的形态，导致创新思维受限制。例如，要求你用六根火柴在桌子上构造四个三角形，由于受桌面结构的影响，而从平面图形的结构上进行思维，总是行不通。若能跳出桌面结构的局限，沿着空间结构进行思维就会恍然大悟，原来六根火柴就可构造具有四个三角形的空间四面体。

(4)信息障碍

在当今时代，信息的影响十分巨大。例如技术情报、专利信息、网络信息等。平时应经常查阅有关信息、资料，以免消息封闭，跟不上时代的步伐。

## 3.3　创新思维　Innovative thought

创新思维是一个哲学命题，有别于常规思维。创新思维是人类思维的最高形式，反映事物本质和内在、外在的联系，是有独特见解的思维过程。它不是单一的思维形式，而是以各种智力与非智力因素为基础，在创新活动中表现出来的具有独创的、产生新成果的高级与复杂的思维活动，是整个创新活动的实质和核心。创新思维的实质或外在表现形式为“选择”“突破”“重建结构”。

我们可对思维形成作如下两个定义：

1)常规性思维(再生思维)。人类在生产与生活实践中碰到的问题能够运用已知的经验、理论和方法解决。虽然也要进行思维，但其新颖性和独特性较差。

2)创新思维(创造性思维)。碰到的问题较复杂，不能直接依靠先前已经掌握的惯常经验、知识、理论、方法等解决时，必须经过独立思考，将储存在头脑中的各种信息重新分析和组合，形成新联系，才能满足需要。

创新思维、发明创造并不是神秘莫测和高不可攀的。现代医学表明，人脑的左半部分擅长抽象思维、分析、数学、语言、意识活动；右半部分擅长于幻想、想象、色觉、音乐、韵律等形象思维和辨认、情绪活动。但人脑的左右部分有两亿条左右的神经纤维相连，形成一个网状结构的神经纤维组织，接收与处理人脑各区域已经加工过的信息，使创新思维成为可能。因此说思考是人类最根本的资源，创新能力人皆有之，思维也是一种可以后天训练培养的技能，通过训练人们能更有效地运用自己的思维，发挥其潜能，使常规思维升华为创新思维。另外，高智商不一定伴随很全面的思维技能，仅表示此人聪明，并不代表他具有好的创新思维能力。发明创造也有其偶然性，不一定与发明人所从事的工作有关。如轮胎的发明人是一个兽医，安全剃须刀的发明者是一个音乐家……

创新思维是人类思维的高级阶段，它是发散思维、收敛思维、直觉思维、灵感思维等多种形式的协调统一，是高效综合运用反复辩证发展的过程，而且与情感、意志、创新动机、理想信念、个性等非智力因素密切相关，是智力与非智力因素的和谐统一。把握创新思维的关键是在认识不同思维的特点、功用的基础上，把它们用到该用的地方。

### 3.3.1 创新理念 Innovative idea

创新与创造是目前使用频度较高的词汇，两者实质上基本相同，只不过创新一词适应的领域更广一些。因此，人们逐渐习惯用创新一词。就其一般意义而言，创新是人类追求新颖、独特并有价值的产物的活动，这里的"产物"既可以是一种新观念、新思想、新理论、新设计，也可以是一种新产品、新工艺、新方法。简而言之，创新就是求新、求异、求奇、求合理。创新也可以具体到以下五个方面：①引入一种新产品或者赋予产品一种新的特性；②引入一种新的生产方法；③开辟一个新的市场；④获得一种新的原材料或半成品新的供应来源；⑤实现一种新的工业组织。

对不同行业而言，创新可划分为两大类型：制度创新和科技创新。两者关系：制度创新为企业发展提供有效的机制保证，为企业带来活力；而技术创新则为企业的发展提供手段，为企业带来竞争力。只有两者的有机组合才能实现企业持续跳跃发展，使企业立于不败之地。

科技创新不同于技术发明。发明是一种创新，但创新绝不仅仅是发明。如果说发明是在新知识、新理论创造基础上一种全新技术出现的话，那么创新则既可能是这种全新技术的开发，也可能是原有技术的改善，甚至可能仅是原有技术的一种简单的重新组合。美国管理学者德鲁克曾以集装箱的生产为例，指出"把车身从车轮上取下，放到货船上"。在这个概念中并没有包含多少技术，可这是一项创新。这项创新缩短了货船留港的时间，把远洋货船的生产率提高了三倍左右，节省了运费，使过去40年中世界贸易得到迅猛的扩大。

### 3.3.2 创新性思维的特征 Characteristics of innovation thought

创造学理论认为，创新性思维除了具有思维的一般属性外，还具有一些它自己的特征。

1.求异性

求异性主要是针对求同性而论的。求同性是人云亦云，照葫芦画瓢。而求异性则是与众人、前人不同，是独具卓实的思维。

求异性思维强调思维的独特性。其思维角度、思维方法和思维路线别具一格，标新立异，对权威与经典敢怀疑，敢挑战，敢超越。求异性思维强调思维的新颖性。其表现为，提出的问题独具新意，思考问题别出心裁，解决问题独辟路径。新颖性是创新行为的最宝贵的性质之一。例如，美国某公司的一位董事长有一次在郊外看一群孩子玩一只外形丑陋的昆虫，爱不释手。这位董事长当时就想，市面上销售的玩具一般都形象俊美，假如生产一些形状丑陋的玩具，效果又会如何呢？于是他安排自己的公司研制了一套"丑陋玩具"，迅速推向市场。结果一炮打响，丑陋玩具深受孩子们的青睐，非常畅销，给该公司带来巨大的经济效益。

2. 突发性

突发性主要体现在直觉与灵感上。所谓直觉思维是指人们对事物不经过反复思考和逐步分析,而对问题的答案做出合理的猜测、设想,是一种思维的闪念,是一种直接的洞察。灵感思维也常常是以一闪念的形式出现的,但它不同于直觉,灵感思维是由人们的潜意识与显意识多次叠加而形成的,是长期创造性思维活动达到的一个必然阶段。

例如,伦琴发现X射线的过程就是一个典型的实例。当时,伦琴和往常一样在做一个原定实验的准备,该实验要求不能漏光。正当他一切准备就绪开始实验时,突然发现附近的一个工作台上发出微弱的荧光,室内一片黑暗,荧光从何而来呢?此时,伦琴迷惑不解,但又转念一想,这是否是一种新的现象呢?他急忙划一根火柴来看一个究竟,原来荧光发自一块涂有氰亚铂酸钡的纸屏。伦琴断开电流,荧光消失,接通电流,荧光又出现了。他将书放到放电管与纸屏之间进行阻隔,但纸屏照样发光。看到这种情况,伦琴极为兴奋,因为他知道,普通的阴极射线是不会有这样大的穿透力的,可以断言肯定是一种人所未知的穿透力极强的射线。经过40多天的研究、试验,终于肯定了这种射线的存在,还发现了这种射线的许多特有性质,并且将其命名为X射线。

事实上,在伦琴发现X射线之前,就曾有人碰到过这种射线,他们不是视而不见,就是因干扰了其原定的实验进行而气恼,结果均失掉了良机。而伦琴则不同,他抓住了突发的机遇,追根溯源,终于取得了伟大的成功。

3. 灵活性

创造性思维的方式、方法、程序、途径等都没有固定的框架,进行创造性思维活动的人在考虑问题时可以迅速地从一个思路转向另一个思路,从一种意境进入另一种意境,多方位地试探解决问题的办法。例如面对一个处于世界经济趋于一体化、竞争日趋激烈之中的小企业的前途问题时,企业的职业经理不能无动于衷或沿用老思路,否则,只有死路一条。企业职业经理必须或是考虑引进外资,联合办厂,或是改组企业的人力、财力、物力的配置结构,并进行技术革新,或加强产品宣传,并在包装上下工夫,或是上述三者并用。企业职业经理也可以考虑企业的转产,或者让某一大型企业兼并,成为大企业的一个分厂。这两种不同思路的创新都是创造性思维在拯救该企业问题时的应用。

如我们熟知的一个例子:20世纪50年代,A国和B国的两家皮鞋公司各派了一名推销员到太平洋一个岛屿去推销皮鞋。两名推销员到岛上一看,该岛居民都没有穿鞋子的习惯。B国的推销员想也不想就打道回府了。A国的推销员转念一想:岛民没有穿鞋的习惯,正好是推销皮鞋的潜在市场。于是向公司要求在岛上多待几天,以调查岛民没有穿鞋的真正原因和思考创造岛民乐于穿鞋的方案。一年后,B国人发现A国人在岛上开了家皮鞋公司,创造了岛民穿鞋的新习惯,并垄断了该岛的皮鞋经营业务。B国人为自己没有创造性后悔莫及。

4. 联想性

爱国斯坦说过,想象比知识更重要,因为知识是有限的,而想象力概括着世界上的一切,推动着进步并且是知识进化的源泉。严格地说,想象力是科学研究中的实在因素。列宁说过,有人认为,只有诗人才需要幻想。这是没有理由的,而且是愚蠢的偏见!甚至在数学上也是需要幻想的,没有它就不可能发明微积分。

5. 反常规性

创新性思维往往以违反常情和不合逻辑的形式出现,因而它也常常不易被人理解。

1938年，一个叫拜罗的匈牙利人发明了活塞式笔芯的圆珠笔，这种“拜罗笔”一度风行世界，但它有个漏油的缺点，到20世纪40年代就几乎被消费者所抛弃。1945年，美国企业家雷诺兹发明了靠重力输送油墨的圆珠笔，但油墨外漏的难题仍未解决。人们为解决难题，研究发现漏油是圆珠磨损变小后产生的，就想方设法更换圆珠的材料，但结果都不满意。1950年，日本发明家中田藤三郎细心研究发现：圆珠笔写到大约2万个字时才开始漏油。于是他一反通常的做法，控制圆珠笔中的油量，使它写到1.5万个字时油刚好用完，笔便弃之不用了。这一下子就解决了这个久久未能解决的难题，为这种新型书写工具的大量使用扫除了消费者的忧虑。

美国宣传奇才哈利十五六岁时，在一家马戏团做童工，负责在马戏场内叫卖小食品。但每次看的人不多，买东西吃的人更少，尤其是饮料，很少有人问津。有一天，哈利的脑袋里诞生了一个想法：向每一个买票的人赠送一包花生，借以吸引观众。但老板不同意这个“荒唐的想法”。哈利用自己微薄的工资作担保，恳求老板让他试一试，并承诺说，如果赔钱就从工资里扣，如果赢利自己只拿一半。于是，以后的马戏团演出场地外就多了一个义务宣传员的声音：“来看马戏，买一张票送一包好吃的花生！”在哈利不停的叫喊声中，观众比往常多了几倍。观众们进场后，小哈利就开始叫卖起柠檬冰等饮料。而绝大多数观众在吃完花生后觉得口干时都会买上一杯，一场马戏下来，营业额比以往增加了十几倍。

6.顿悟性

创新性思维是在人们苦思冥想之后，以一种突然的形式在人们头脑中闪现的。

阿基米德发现浮力定律的传说能够很好地证实这一点。希腊国王请艺匠用纯金打了一顶王冠，王冠打好后，国王觉得不太像是纯金的，可是又没有办法证实这一点。他请阿基米德来做这一鉴定工作，而且要求不破坏王冠本身，因为并不能肯定其中掺有银子，要是把王冠毁坏了而其中又没有掺假，那代价就大了。阿基米德一直在思考这一问题，但没有找到较好的鉴定方法。有一天，他正准备进浴盆里洗澡，这一次仆人把水放得太满，当他坐进浴盆时有许多水溢了出来。这使他一下子想到溢出的水的体积正好应该等于他自身的体积，如果他把王冠浸在水中，根据水面上升的情况可以知道王冠的体积，拿与王冠同等质量的金子放在水里浸一下，就可以知道它的体积是否与王冠体积相等，如果王冠体积更大，就说明其中掺了假。阿基米德想到这里，十分激动，他一下从浴盆里跳了起来，光着身子就跑了出去，一边跑还一边喊：“尤里卡”，喊出了人类探寻到大自然奥秘时的惊喜，表达了创造性思维的顿悟性特征。为了纪念这一事件，现代世界最著名的发明博览会以“尤里卡”命名。

7.风险性

由于创造性思维活动是一种探索未知的活动，因此要受到多种因素的限制和影响，如事物发展及其本质暴露的程度、实践的条件与水平、认识的水平与能力等，这就决定了创造性思维并不能每次都取得成功，甚至有可能毫无成效或者得出错误的结论。例如，西欧中世纪，宗教在社会生活中占据着绝对统治地位，一切与宗教相悖的观点都被称为“异端邪说”，一切违背此原则的人都会受到“宗教裁判所”的严厉惩罚。但是，创造性思维活动是扼杀不了的，伽利略、布鲁诺置生命于不顾，提倡并论证了“日心说”，证明教皇生活于其上的地球不是宇宙的中心。无法想象，如果没有两位科学家甘愿冒此风险，“日心说”不知何时被提出。

8.可迁移性

创造力开发的实践证明，从一种情境开发的创新性思维能力，可以迁移到其他情境中去，这是我们进行创新性思维训练、开发人的创造力的理论依据。

### 3.3.3 创新思维的作用 Roles of innovation thought

首先,创造性思维可以不断地增加人类知识的总量,不断推进人类认识世界的水平。创造性思维因其对象的潜在特征,表明它是向着未知或不完全可知的领域进军,不断扩大着人们的认识范围,不断地把未被认识的东西变为可以认识和已经认识的东西,科学上每一次的发现和创造,都增加着人类的知识总量,为人类由必然王国进入自由王国不断地创造着条件。

其次,创造性思维可以不断地提高人类的认识能力。创造性思维的特征已表明,创造性思维是一种高超的艺术,创造性思维活动及过程中的内在的东西是无法模仿的。这内在的东西即创造性思维能力。这种能力的获得依赖于人们对历史和现状的深刻了解,依赖于敏锐的观察能力和分析问题能力,依赖于平时知识的积累和知识面的拓展。而每一次创造性思维过程就是一次锻炼思维能力的过程,因为要想获得对未知世界的认识,人们就要不断地探索前人没有采用过的思维方法、思考角度去进行思维,就要寻求独创性的办法和途径去正确、有效地观察问题,分析问题和解决问题,从而极大地提高人类认识未知事物的能力,因此,认识能力的提高离不开创造性思维。

再次,创造性思维可以为实践开辟新的局面。创造性思维的独创性与风险性特征赋予了它敢于探索和创新的精神,在这种精神的支配下,人们不满于现状,不满于已有的知识和经验,总是力图探索客观世界中还未被认识的本质和规律,并以此为指导,进行开拓性的实践,开辟出人类实践活动的新领域。在中国,正是邓小平创造性的思维,提出了有中国特色的社会主义理论,才有了中国翻天覆地的变化,才有了今天轰轰烈烈的改革实践。相反,若没有创造性的思维,人类躺在已有的知识和经验上,坐享其成,那么,人类的实践活动只能留在原有的水平上,实践活动的领域也非常狭小。

最后,创造性思维是将来人类的主要活动方式和内容。历史上曾经发生过的工业革命没有完全把人从体力劳动中解放出来,而目前世界范围内的新技术革命带来了生产的变革,全面的自动化把人从机械劳动和机器中解放出来,从事着控制信息、编制程序的脑力劳动,而人工智能技术的推广和应用,使人所从事的一些简单的、具有一定逻辑规则的思维活动,可以交给“人工智能”去完成,从而又部分地把人从简单脑力劳动中解放出来。这样,人将有充分的精力把自己的知识、智力用于创造性的思维活动,把人类的文明推向一个新的高度。

### 3.3.4 创新思维常用形式 Forms of innovative thought

1)形象思维。它是对实物的映像、图形、符号、模型、形体等不同形式构成所谓的形象进行思考的思维方式。运用形象思维可以激发人们的想象力、联想和类比能力。

例如行星轧压法的发明,以往将金属轧制成板材,是将金属原料送到两个轧辊之间,靠两个轧辊的转动和原材料的运动而完成的。这种方法对于延展性能良好的钢材是适用的,但对于延展性能差的材料,在轧制中会出现裂纹。为了解决这个问题,日本某特种钢厂的一位技术人员绞尽脑汁。一天,当他去厨房喝水时,无意中被妻子在面板上擀面的姿势和方法所吸引,于是想出了轧制钢材的一种新方法——行星轧压法。

又如我们熟知的一个例子:爱因斯坦提出相对论后,一天一位不知相对论为何物的年轻人向爱因斯坦请教相对论。相对论是爱因斯坦创立的既高深又抽象的物理理论,要在几分钟内让一个门外汉弄懂什么是相对论，简直比登天还难。然而爱因斯坦却用十分简洁形象的话对深奥的相对论给出了解释："比方说，你同亲爱的人在一起聊天，一个钟头过去了，你只觉得过了五分钟；可如果让你一个人在大热天孤单地坐在炽热的火炉旁，五分钟就好像一个小时。这就是相对论！"

形象思维在每个人的思维活动和人类所有实践活动中，均广泛存在，具有普遍性。许多设计，许多科学的发明创造，往往是形象思维中受到启发而产生的，有时还会取得抽象思维难以取得的成果。钱学森认为："人们对抽象思维的研究成果曾经大大地推进了科学文化的发展"，那么"我们一旦掌握了形象思维，会不会用它来掀起又一次新的技术革命呢？这是值得玩味的设想"。

2)抽象思维。抽象思维又称逻辑思维，是以认识过程中反映事物共同属性和本质属性的概念作为基本思维形式，在概念的基础上进行判断、推理，是反映事实的一种思维方式，使认识由感性个别到理性一般再到理性个别。一切科学的抽象，都会更深刻、更正确、更完全地反映客观事物的面貌。

归纳和演绎、分析和综合、抽象和具体等，是抽象思维中常用的方法。所谓归纳的方法，即从特殊、个别事实抽象到一般概念、原理的方法；而演绎的方法，则是从一般概念、原理推出特殊、个别结论的方法；所谓分析的方法，是在思想中把事物分解为各个属性、部分、方面分别加以研究；而综合则是在头脑中把事物的各个属性、部分、方面结合成整体。作为思维方法的抽象，是指由感性具体到理性抽象的方法，具体则指由理性抽象到理性具体的方法。它们是相互依存、相互促进、相互转化的，彼此相反而又相互联系。

例如元素周期表的发现，当时大多数科学家均热衷于研究物质的化学成分，醉心于发现新元素，无人去探索化学中的"哲学原理"。而门捷列夫却在寻求庞杂的化合物、元素间的相互关系，寻求能反应内在本质属性的规律。他不但把所有的化学元素按原子量的递增及化学性质的变化排成合乎自然规律、具有内在联系的一个个周期，而且还在表中留下了空位，预言了这些空位中的新元素，也大胆地修改了某些当时已经公认了的化学元素的相对原子质量。

3)发散思维。发散思维又称求异思维或辐射思维。它不受现有知识和传统观念的局限与束缚，是沿着不同方向多角度、多层次去思考、去探索的思维形式。流畅、灵活和独特是发散思维三个不同层次的特征。流畅性表明能在短时间内表达出较多的概念、想法，表现为发散的"个数"指标。灵活性指在思维时能触类旁通，随机应变，不受心理定势影响，能多方面、多角度进行思考，表现为发散的"类别"指标。独特性是更高层次，即能提出超乎寻常的观念，表现为发散的"新异、独特"指标。

例如测试发散思维的层次性可用如下方式：

题目：要求被测试者在5分钟内列出砖的可能用途（举出一种用途得1分，举出一种类别得1分，有独特性得1分）。

甲答案：造房、铺路、砌灶、造桥、保暖、堵洞、做三合土、填物。

得分：流畅性得8分，灵活性1分，因为全是一种材料、无独特性，共计9分。

乙答案：造房、铺路、防身、敲击、压纸、量具、积木玩具、耍杂技、磨成粉来当颜料。

得分：流畅性9分，灵活性6分，因类别为材料、武器、工具、量具、玩具、颜料，独特性2分，共计17分。说明乙比甲的发散思维水平高。

4)收敛思维。收敛思维亦称集中思维、求同思维或定向思维，是以某个思考对象为中心，从不同角度、不同方面将思路指向该对象，以寻找解决问题的最佳答案的思维形式。在设想的实现阶段，这种思维形式常占主导地位。

在创造性思维过程中，发散与收敛思维是相辅相成的。只有把二者很好地结合使用，才能获得创造性成果。美国哲学家库恩认为："科学只能在发散与收敛这两种思维方式相互拉扯所形成的张力之下向前发展。如果一个科学家具有在发散式思维与收敛式思维之间保持一种必要的张力的能力，那么他就具备从事科学研究所必需的条件之一。"

举一个病人去医院看病的简单例子：病人向医生诉说常常低热不退。这仅仅是一个"症状"。究竟是什么原因引起此症状的呢？医生常用的即是发散思维的方法——可能是体内炎症，可能是肺结核，可能是神经官能症，还可能是癌症……医生就要继续询问各种病症，并做必要的检查、化验。待病因确诊后，就用收敛思维的方法，用一切可行方案集中力量将病治好。

5)灵感思维。灵感是人们借助于直觉启示而对问题得到突如其来的领悟或理解的一种思维形式，是一种把事物信息隐藏在潜意识中，当需要解决某个问题时，其信息就以适当的形式突然表现出来的创造能力。它是创造性思维最重要的形式之一。有人称灵感是创造学、思维学、心理学皇冠上的一颗明珠，是很有道理的。

科学也已证明，灵感不是玄学而是人脑的功能，在大脑皮层中有对应的功能区域，即由意识部和潜意识部两个对应组织所构成的灵感区。意识部和潜意识部相互间的同步共振活动主导灵感的发生。灵感的产生亦需一定的诱发因素，有其客观的发生过程，是偶然性与必然性的统一。灵感的出现不管在空间上还是在时间上都具有不确定性，但灵感的产生条件却是相对确定的。它的出现有赖于知识长期的积累，有赖于智力水平的提高，有赖于良好的精神状态与和谐的外界环境，有赖于长时间、紧张的思考和专心的探索。

法国数学家热克·阿达马尔把灵感的产生分为准备、潜伏、顿悟和检验四个阶段；也有人把其分为准备期、酝酿期、豁朗期和验证期。这两者是相一致的。准备与潜伏期，是长期积累、刻意追求、循常思索的阶段；顿悟是由主体的积极活动和过去的经验所准备的、有意识的瞬时的动作，是思维过程中逻辑的中断和思想的升华，是偶然得之、无意得之、反常得之的顿发思索阶段。当灵感顿发时，往往会伴随着一种亢奋性的精神状态。可以把灵感分为来自外界的偶然机遇型与来自内部的积淀意识型两大类。

在各类创造性灵感中，由外部偶然的机遇而引发的灵感最为常见、有效。有人说："机遇，是发明家的上帝。"这是极有道理的。

过去挖藕的方法，均是在天冷时由人用耙子下到水中去挖，又脏又累。有一次，一人挖藕时放了一个屁，众人大笑，但其中一人却马上想到：如果将压缩空气吹入池底，是否可挖藕？经试验，将水加压后喷入池底，莲藕不但被挖出，而且又干净且不损坏。于是，一种新的挖藕的方法得到普遍采用。

工业设计师、建筑师设计时，从自然界各种形态中得到灵感，产生出许多优秀设计的实例，更不胜枚举。

6)逆向思维。逆向思维也称反向思维，即把原思维反向逆转，用与原来的想法对立的或表面上看似不可能并存的两条思路去寻求问题答案的思维方式。

常用的数字运算，是从低位到高位运算，而"快速计算法"却从高位向低位运算；市场销售中的反季节销售也是逆向思维的结果。

反向思维在突破传统观念，提出富有创造性的设想或方案方面是很有成效的。运用反向思维，首先要找到传统思路，然后按照与它背道而驰的思路去思考，寻求新的创见。

下面是几种常用的反向思维方式：

• 时序逆反：从时间顺序上进行反向思维。例如反季节销售的营销方法和反季节的蔬菜种植。

• 功用逆反：从事物的功用方式上进行反向思维。例如手表的后盖通常是用不锈钢等不透明材料制成的，给人感觉上是在保护机芯。但有人反过来思考，用透明材料制成后盖，能满足人们的好奇心，使其直接观察到机芯的运动情况。

• 结构逆反：从事物的结构方面进行反向思维。例如热源上置烤肉方法，减少了油烟，降低了环境污染。

• 原理逆反：针对事物的基本原理进行反向思维。例如物理学家法拉第在懂得电磁效应规律后，反过来思考：磁会不会有电效应？经过长期探索，终于发现了电磁感应现象，并建立了电磁感应定律。

7)灵感思维。所谓灵感，是指人们在久思某一问题不得其解时思绪由于受到某种外来信息的刺激或诱导，忽然灵机一动，想出了办法，对问题的解决产生重大领悟的思维过程。

以发明袖珍电脑和袖珍电视而文明的英国发明家辛克莱为例，他在谈到怎样设计出袖珍电视时，曾这样写道："我多年来一直在想，怎样才能把显像管的长尾巴去掉。有一天，我忽然来了灵感，巧妙地把尾巴做成了九十度弯曲，使它从侧面而不是从后面发射电子，结果就设计出了厚度只有3厘米的袖珍电视机。"

8)联想思维。联想思维就是根据当前知道的事物、概念或现象，联想到与之相关的事物、概念或现象的思维活动。由于客观事物之间的联系是多种多样的，不同人大脑搜寻记忆库的方式也不同，因此，联想就出现了多种类型。

• 相关联想：由给定事物联想到经常与之同时出现或在某个方面有内在联系的事物的思维活动。例如由鸡可联想到鸡蛋，也可联想到鸡类，还有人联想到"鸡窝里飞出个金凤凰"等，过些都是相关联想。

• 相似联想：是从给定事物想到与之相似(形状、位置、性质等方面)的事物的思维活动。例如：烧饼——十五的月亮(形状相似)；飞鸟——飞机(功能相似)；香味——花香(气味属性相似)。

• 类比联想：是由一类事物的规律或现象联想到其他类事物的规律或现象的思维活动。美国工程师斯潘塞，主要研究波长略长于25cm的电波在空间的分布情况，有一天，他正在做雷达起振实验，忽然同事们惊叫起来："你受伤了，出血了！"果然暗红色的血迹，从上衣口袋里渗出来。他手一摸，湿乎乎的，不禁大惊失色。可是他立刻又明白过来，那渗出来的只是熔化了的巧克力糖液。这一现象非常奇怪，周围没有炉子，温度也远不能令巧克力熔化，那么巧克力到底为什么会熔化呢？答案只有一个：雷达发出的强大电波——微波——使巧克力内部分子发生振荡，从而产生了热能，斯潘塞由此联想到：微波可以使巧克力熔化，那么同样可以使其他食品内外同时受热，可以利用微波把食物"烧熟"，而且这样"烤"非常均匀。经过研制，斯潘塞发明了微波炉。

• 对称联想：是由给定事物联想到在空间、时间、形状、特性等方面与之对称的事物的思维活动。例如：光明——黑暗、放大——缩小、船——潜艇。

1901年出现的除尘器是吹式的，当在伦敦某火车站一节车厢里演示时，这种除尘器曾将车厢里吹得尘土飞扬，叫人透不过气来。这个现象引起了一位名叫赫伯布斯的在场者的注意。

他想："吹尘不行，那么反过来吸尘行不行呢？"回家后，他用手绢捂住嘴，趴在地毯上使劲吸气，结果灰尘被吸滤到手绢上了。赫伯布斯因此发明了带灰尘过滤装置的真空负压吸尘器。

将联想思维分成若干类型，有助于人们认识联想的创新能力，但在创新过程中不能预先规定好使用哪种类型的联想去帮助创新，只能是让联想去自由地发挥。创新往往是多种联想的综合结果。

9)直觉思维。直觉是人类一种独特的"智慧视力"，是能动地了解事物对象的思维的闪念。直觉思维能以少量的本质性现象为媒介，直接把握事物的本质与规律，是一种不加论证的判断力，是思想的自由创造。

创新性思维的本质，表现为选择、突破和重新建构。而要做出选择，无疑取决于人们直觉能力的高低，美籍华裔物理学家丁肇中在谈到J粒子的发现时写到："1972年，我感到很可能存在许多有光的特性而且有比较重质量的粒子。然而理论上没有预言这些的存在。我直观上感到没有理由认为重光子一定比质子轻。"正是在这种直觉的驱使下，丁肇中决定研究重光子，终于发现了J粒子，并因此获得了诺贝尔物理学奖。

伟大的科学家爱因斯坦认为："真正可贵的因素是直觉。"同时，他认为科学创造原理可简洁表达成：经验—直觉—概念—逻辑推理—理论。他说："我相信直觉和灵感。"苏联科学史专家凯德洛夫指出："没有任何一个创造性行为能够脱离直觉活动。"可见直觉的重要性。当然，直觉思维也有其自身的缺点。例如，容易把思路局限于较狭窄的观察范围里，会影响直觉判断的正确性和有效性，也可能会将两个本不相关的事纳入虚假的联系之中，个人主观色彩较重。因此，关键在于创新者主体素质的加强和必要的创新心态的确立，而且还必须有一个实践检验过程，这是重要的科学创新阶段。

### 3.3.5 创新思维的过程 Process of innovation thought

创新思维是一个复杂的思维过程，许多心理学家对这个过程进行了研究，都认为这个过程可分为若干个阶段，但在阶段的划分上却又各持己见，我们将其归纳成如下几类：

1)三阶段说。约翰逊1955年提出创造性思维可分为准备、产物和判断三个阶段。彭加勒也曾将数学创造的思维过程划分为意识的逻辑分析、潜意识活动以及潜意识活动的产物转化为意识这样3个阶段。美国技术预测学家捷恩茨还提出了一个包括13个步骤的创造性思维过程三阶段模式（见表3-1）。

**表3-1　创造性思维过程三阶段模式**

| 阶段 | 步骤 | 名称 | 思维形式 |
| --- | --- | --- | --- |
| 一 | 1 | 前导 | U/C |
| | 2 | 不满 | U/C |
| | 3 | 认识环境 | C |
| | 4 | 获得资料 | C |
| | 5 | 研究分析 | C |
| 二 | 6 | 潜伏 | U |
| | 7 | 顿悟 | U |

续表

| 阶段 | 步骤 | 名称 | 思维形式 |
| --- | --- | --- | --- |
| 三 | 8 | 产生 | U |
| | 9 | 发展 | C |
| | 10 | 审核 | C |
| | 11 | 实施 | C |
| | 12 | 满意 | U/C |
| | 13 | 转向 | U/C |

2)四阶段说。美国心理学家沃拉斯最早提出将创造性思维过程分为准备、酝酿、明朗、证验等四个阶段。这种划分对心理学界的影响极大,现已成为创造学界广为接受的形式。

3)五阶段说。莫瑞非尔德曾提出将创造性思维过程划分为准备、分析、产物、证验和应用五个阶段的观点。苏联的戈加内夫和英国的劳森都提出过类似的划分。美国的杜威在其名著《我们是怎样思维的》一书中将五个阶段描述为①感到某种困难的存在;②认清是什么问题;③搜集资料,进行分类,并提出假说;④接受或抛弃试验性假说;⑤得出结论并加以评论。

4)七阶段说。美国创造美学家斯本把创造性思维过程划分成七个阶段:①定向,强调某个问题;②准备,收集有关资料;③分析,把有关材料进行分类;④观念,用观念进行各种各样的选择;⑤沉思,“松弛”,促使启迪;⑥综合,把各个部分结合在一起;⑦估价,判断所得到的思想成果。美国现代心理学家罗斯门也有类似的划分。

对创造性思维划分阶段,有利于我们把握其关键,正确掌握其思维规律,并有目的地进行创造性思维训练,提高创造力。我们也比较赞同四阶段说。下面就对创造性思维过程的四个阶段进行更为详细的讨论。

1)准备阶段。创新思维是从发现问题、提出问题开始的。“问题意识”是创新思维的关键,提出问题后必须为解决问题做充分的准备。这种准备包括必要的事实和资料的收集,必需的知识和经验的储备,技术和设备的筹集以及其他条件的提供等。同时,必须对前人在同一问题上所积累的经验有所了解,并对前人在该问题尚未解决的方面作深入的分析。这样既可以避免重复前人的劳动,又可以使自己站在新的起点从事创造工作,还可以帮助自己从旧问题中发现新问题,从前人的经验中获得有益的启示。准备阶段常常要经历相当长的时间。

例如,爱因斯坦青年时期就为物理学中的基本问题感到不安,尤其是光速问题,他日夜思考,长达 7 年之久。后来当他突然想到解决方案时,只花了 5 周时间就写出了闻名世界的《相对论》论文。人们以为爱因斯坦的这一创造性思维只有 5 周时间,其实他已花了 7 年时间作准备。在准备阶段,创造性思维的活动主要集中在发现问题,分析问题,形成有创造价值的课题。发现问题是起点,分析问题并形成创造课题是关键。

2)酝酿阶段。酝酿阶段要对前一阶段所获得的各种资料和事实进行消化吸收,从而明确问题的关键所在,并提出解决问题的各种假设和方案。此时,有些问题虽然经过反复思考、酝酿,仍未获得完美的解决,思维常常出现“中断”、想不下去的现象。这些问题仍会不时地出现在人们的头脑中,甚至转化为潜意识,这样就为第三阶段(顿悟阶段)打下了基础。许多人常常在这一阶段表现为狂热和如痴如醉，令常人难以理解。如我们非常熟悉的牛顿把手表当鸡蛋煮、安倍不认识家门、陈景润在马路上与电线杆相撞等故事。

这个阶段可能是短暂的，也可能是漫长的，有时甚至延续好多年。创新者的观念仿佛是在“冬眠”，等待着“复苏”和“醒悟”。

3)顿悟阶段。顿悟阶段也叫做豁朗阶段，经过酝酿阶段对问题的长期思考，创新观念可能突然出现，思考者大有豁然开朗的感觉。

灵感的来临，往往是突然的、不期而至的。如德国数学家高斯，为证明某个定理，被折磨了两年仍一无所得，可是有一天，正如他自己后来所说，“像闪电一样，谜一下解开了”。

4)验证阶段。思路豁然贯通以后，所得到的解决问题的构想和方案还必须在理论上和实践上进行反复论证和试验，验证其可行性。经验证后，有时方案得到确认，有时方案得到改进，有时甚至完全被否定，再回到酝酿期。总之，灵感所获得的构想必须经过检验。

下面，我们通过两个事例来具体地了解创造性思维过程的上述四个阶段。

例1:1974年，周林就读于上海交通大学船舶电气自动化系。每年冬天，他和许多人一样手脚长满冻疮，痒痛难忍，四处求医用药，都无济于事。冻疮的痛苦折磨着他，也引发了他的思考:“难道世界上就没有更好的办法治疗冻疮吗?”发现了问题，周林开始分析问题，他跑到图书馆查阅资料。周林在有关资料中发现早在公元前230年我国就有人为坚守边关将士研究治疗冻疮的方法。之后，历代名医学者一直在进行研究，直到现代医学，虽然提出了很多种治疗方法，但效果都不尽如人意。在中国绝大部分地区都可发生冻疮，每年冬季至少有3 000万人遭受冻疮的折磨。在国外，两次世界大战中，参战国因为冻疮的危害，非战斗减员上百万。这个问题的研究，已成为许多国家关注的问题。周林震惊了，他决心利用业余时间摸摸冻疮这只老虎的屁股！于是，创造课题就形成了。

形成创造课题后，周林开始搜集资料进行研究活动。他利用各种机会收集民间偏方进行试验:把冬天的雪水用罐封住，第二年夏天用来浸手脚；把西藏传统的木炭烧红后扔在盆里加热，再把冻坏的手脚放在热水中洗烫；还用萝卜煎水清洗手脚；用酒精胡椒酊剂清洗手脚。各种方法试来试去，结果都不理想。

1977年，周林大学毕业回到云南工作，在紧张的工作之余仍冥思苦想治疗冻疮的问题。他综合多学科的技术知识，对冻伤治疗的现状进行深层的了解。在对国内外有关寒冷性损伤机制的七种假说和九大治疗方法的剖析中，周林逐渐形成了自己的独特见解:现有的各种疗法，虽然都是针对冻疮病有的放矢，但疗效都欠佳，其原因在于药物没有在冻损区充分发挥药效，一般都是以痛压痒，以热抑痛的治标方案。由此，周林产生了一个大胆的设想:走不用药物治疗冻疮的新路。

这条新路该怎样走呢?周林经过反复思索，决定从生物医学工程方面入手寻求答案。于是，他开始学习有关生物医学工程方面的知识，并进行了必要的试验和测试。在对冻疮患者的测试中，他发现冻疮病人末梢患处的皮肤温度比正常人的同一部位低得多，平均只有22℃。之后，他借助微循环显微仪观察，发现正是由于患处温度低，所以都存在着明显微循环不良的症状。根据这些情况，周林又思索开了:“看来，我只有另辟蹊径，从人体组织内部入手，才有可能找到一种既简便又治本的新方法。”可是，要找到这种新方法谈何容易。在那些日子里，他吃饭在想，走路在想，睡觉也在想。冥思苦想了很久，还是“没有结果”，这时，周林来了个暂时中断思索，将问题搁置一旁，去考虑其他工作上的问题，但是潜意识仍在活动。

直到有一天，周林在车间劳动，站在一台大型砂轮机旁边帮助工人打磨铸件。突然，“哒哒哒”，沉重的铸件在砂轮的磨削下产生了巨大的震动，震得他浑身颤抖，一股强大的振荡冲击波从双手传到大脑，使他热血沸腾。瞬间，一个灿烂的创造火花在他脑海中闪现：“振动?！谐振?！匹配作用?！我现在感到全身血液流通发热，不是刚才的谐振所引起的吗？如果将谐振原理与冻疮治疗建立起联系，发明一种治疗仪器该多好啊！”周林顿时豁然开朗，潜伏在头脑中多年的创造灵感一下子迸发出来。经过一段时间的研究试制，周林开发了具有实用价值的冻疮治疗仪。1985年10月，周林因冻疮治疗仪的发明荣获首届世界青年发明家科技成果展览会金牌奖。在那次会上，周林发明的冻疮治疗仪大显神通，大会专刊以醒目标题称赞此项发明是“全世界都欢迎的治疗冻疮的有效武器”，是“具有东方魔力的神奇仪器”。

例2：莱特兄弟少年时，爸爸从外地回来给他们带了一份圣诞礼物，兄弟俩打开一看，是一个不知名的玩具，样子非常奇怪。爸爸告诉他们，这是飞螺旋，能在空中高高地飞翔。“鸟才能飞呢！它怎么也会飞?！”维尔伯有点怀疑。爸爸笑了笑，当场做了表演。只见他先把上面的橡皮筋扭好，一松手，它就发出呜呜的声音，向空中高高地飞去。兄弟俩这才相信，除了鸟、蝴蝶之外，人工制造的东西，也可以飞上天。于是，兄弟俩便把它拆开了，想从中探索一下，它为何能飞上天去。从这以后，在他们的幼小心灵里，就萌发了将来一定要制造出一种能飞上高高蓝天的东西的愿望。这个愿望一直影响着他们。1896年，莱特兄弟在报纸上看到一条消息：德国的李林塔尔因驾驶滑翔机失事身亡。这个消息对他们的震动很大，弟兄俩决定研究空中飞行。于是，创造课题就形成了。

形成创造课题后，他们开始查阅大量资料。由于莱特兄弟开着一家自行车商店，所以他们一边干活挣钱，一边做研究。三年后，他们掌握了大量有关航空方面的知识并决定仿制一架滑翔机。可是做的结果一直不理想。忽然有一天他们看到老鹰在空中飞行，豁然开朗，如果仿制老鹰的形状制作出滑翔机，那是不是就可以飞起来了呢。这个灵感一出来，他们就埋头开始研究。

他们首先观察老鹰在空中飞行的动作，然后一张又一张地画下来，之后才着手设计滑翔机。1900年10月，莱特兄弟终于制成了他们第一架滑翔机，并把它带到离代顿很远的吉蒂霍克海边进行放飞试验。这次成功了，滑翔机飞了起来，但只有1m多高。于是，兄弟俩在上次制作的基础上，经过多次改进，又制成了一架滑翔机。这次试验，飞行高度一下子达到180m之高。

兄弟俩非常高兴，但并不满足。他们想能否制造一种不用风力也能飞行的机器？兄弟俩反复思考，把有关飞行的资料集中起来，反复研究，始终想不到用什么动力才能把庞大的滑翔机和人运到空中。有一天，车行门前停了一辆汽车，司机向他们借一把工具修理一下汽车的发动机。兄弟俩灵机一动，想着能不能用汽车的发动机来推动飞行呢？

从这以后，兄弟俩围绕发动机动开了脑筋。他们首先测出滑翔机的最大运载能力是90kg，在得到了质量只有70kg的汽油发动机后，兄弟俩便着手研究怎样利用发动机来推动滑翔机飞行。经过无数次的试验，他们终于把发动机安装在滑翔机上，不过是在滑翔机上安上螺旋桨，由发动机来推动螺旋桨旋转，带动滑翔机飞行。

1903年9月，莱特兄弟带着他们装有发动机的滑翔机再次来到吉蒂霍克海边试飞。虽然这次试飞失败了，但他们从中吸取了很多经验。过后不久，他们又连续试飞多次，不是因为螺旋桨出了故障，就是因为发动机出了毛病，或是驾驶技术出了问题。莱特兄弟毫不气馁，仍然坚持试飞。就在这时，一位名叫兰莱的发明家，受美国政府的委托，制造了一架带有汽油发动机的飞机，在试飞中坠入大海。

莱特兄弟得知这个消息后，便前去调查，并从兰莱的失败中吸取了教训，获得了很多经验，他们对飞机的每一部件都做了严格的检查，制订了严格的操作规定，于1903年12月14日，又来到吉蒂霍克进行试飞试验。经过多次试验后，终于，在1903年12月17日这天上午10点钟，兄弟俩的飞机飞行了30m后，稳稳地着陆了。维尔伯冲上前去，激动地扑到刚从飞机里爬出来的弟弟身上，热泪盈眶地喊道："我们成功了！我们成功了！"45min后，维尔伯又飞了一次，飞行距离达到了52m，又过了一段时间，奥维尔又一次飞行，这次飞行了59s，距离达到了255m。

这是人类历史上第一次驾驶飞机飞行成功。1908年，莱特兄弟在政府的支持下，创办了一家飞行公司，同时开办了飞行学校。从这以后，飞机成了人们又一项先进的运输工具。

## 3.4　创新技法　Methods of invention

工业设计从事的是创新设计，是一项创造性活动。创造性活动不仅要依赖创新性思维，同时也要掌握并正确运用创新方法和技巧。创新方法或技法是创新性思维的具体化。

在20世纪初，特别是在第二次世界大战以后，工业产品更新换代加速，创新的内涵及规律成为人们研究的课题，且不断深入。在总结了前人的经验之后，大量被实践证明了是行之有效的创新技法应运而生。目前各国已开发出的创新技法有三四百种。各种创新方法的基本出发点是打破传统思维习惯，克服思维定势和妨碍创新设想的各种消极心理因素，充分发挥各种积极心理，以提高创新能力为宗旨。在此我们介绍几种典型技法，读者可详加体会，然后触类旁通，逐渐掌握较多的技法。掌握这些方法，可以使我们的创新活动省力、省时、提高效率，并能使我们多出成果、快出成果。创新技法是一种技巧性的方法，俗话说："熟能生巧。"因此，要掌握创新技法，必须多运用、多实践。

创新本身是一种探索性的社会实践活动，它有规律，但规律不能像自然科学的规律那样可以用数学公式表达，其带有模糊性。人们在这种模糊的规律指导下，尝试着用某种方法去解决创造问题。因此创新技法也具有探索性，它的应用必然因人、因地、因事而异，有时甚至不一定能完全成功和有效。也就是说，创新技法没有万能的，只能针对不同的事情，运用不同的技法。我们应该多学习一些创新技法，以便运用于不同的问题。

日本创造学家还提出了运用创新技法的七原理，表述如下：

1)生动原理。运用创新技法要生动，不要僵化，不要生搬硬套。

2)刺激原理。要自觉地利用各种信息的刺激。

3)希望原理。要通过创新技法的运用，培养起强烈的创新意识和成功欲望。

4)环境原理。要形成良好的创造环境。

5)比较原理。解决同一个问题，可以应用不同的创造技法，有可能的话，要通过比较，选择更有价值的技法和设想。

6)结构原理。

7)概率统计原理。实践证明，创造性设想越多，创造成功的概率也就越大。采用尽可能多的创新技法，获得尽可能多的创造性设想，是提高创造成功率的必要条件。

### 3.4.1 科技创新方法 Methods of technological innovation

所谓科技创新，顾名思义就是涉及科学与技术领域的创新活动，它包括科学发现、技术发明和技术革新三个方面。

1.科学发现

科学发现，就是发现新的科学事实和科学规律，提出新的科学理论的创新活动。

法国科学家未勤德行发现了石灰水和硫酸铜混合液能够防止葡萄露菌病的科学事实，促使人们后来发现了用来防治植物病虫害的“波尔多液”。

古希腊的哲学家亚里士多德根据日常生活中的现象和人们的常识提出了一个著名的原理：当推一个物体的力取消时，原来的物体便归于静止。如果要使物体不停地运动下去，这个外加的力也得持久不断地作用。

16世纪末，意大利物理学家伽利略发现物体的运动并非如亚里士多德所说的那样，而是一个物体一旦不受外力作用，它仍会以一恒定的速度不停地运动下去。即“动者恒动，静者恒静”。这个科学原理的发现为近代力学奠定了坚实的基础。正是在这个基础上，牛顿提出了力学第一定律——惯性定律，进而创立了牛顿力学的理论体系。

还有，达尔文创立进化论，爱因斯坦创立相对论，李四光创立地质力学等等，都属于科学发现。科学发现可以有三个不同的层次：

第一层次，发现科学事实。例如，前面提到的人们发现“波尔多液”，李政道发现J粒子等。

第二层次，在发现一些科学事实的基础上，归纳出科学规律。例如，阿基米德发现“浮力定律”，门捷列夫发现元素周期律都属于这一层次。

第三层次，在归纳出一些科学规律的基础上，创立一整套科学理论。例如，牛顿在归纳力学三定律的基础上，创立了牛顿力学体系，爱因斯坦在归纳出“相对性原理”和“光速不变原理”的基础上，创立了狭义相对论等。

研究一些科学发现现象，我们就可找出科学发现的基本规律，如图3-1所示。

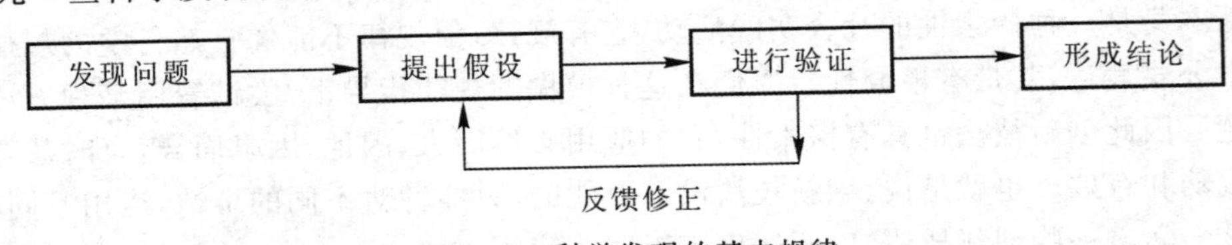


图3-1 科学发现的基本规律

人类通过科学发现，加深对自然界的了解和认识，从辩证唯物主义的观点看，人类对自然界的认识是永无止境的，因而有更多的科学事实、科学规律、科学理论等待着我们去发现、去创立。

2.技术发明

技术发明，是人们运用科学发现的成果创造出一种以前不存在的人工事物或方法的实践活动。技术发明不同于科学发现，但它与科学发现又有着密切的联系。技术发明要运用科学发现的成果，反过来技术发明又会促进科学发现。爱迪生在发明电灯泡的过程中，曾发现了一个重要的物理现象：通电时的灯丝与灯泡内的金属板之间有电流通过。当时并未意识到该发现有什么实用价值，但还是记录在案，并取得了发明专利，后人称之为爱迪生效应。十多年后这个现象就得到了解释：灯丝发热是有电子发射出来，它与金属板之间正好形成回路。英国学者弗莱明于20世纪初利用爱迪生效应发明了电子二极管，引发了另一次技术革命。

技术发明可以分为产品型发明和方法型发明两种类型。产品型发明是指发明人造出来的各种物质性成果，如工具、生活用品、合成物等。方法型发明是指发明人提出的用以制造产品的各种工艺性技术方案，运用这种方案可以使物质状态发生变化。如袁隆平发明的杂交水稻育种法，王永民发明的五笔字型汉字输入法等。技术发明的一般规律如图3-2所示。

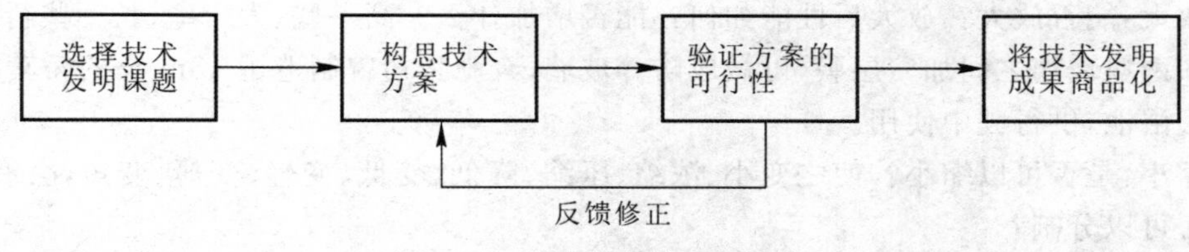


图3-2 技术发明的一般规律

3.技术革新

技术革新，是人们运用科学研究和技术发明的成果，改进已有的人工事物或方法的实践活动。技术革新是技术领域里的创造活动，与科学发现是不相同的。而且，技术革新也不同于技术发明，技术发明是探索性的应用开发研究，是技术开发过程中的初级阶段，是新技术从孕育到产生的创新过程；技术革新是在已有技术发明的基础上的改进与完善，是技术的实用化与工业化。例如，世界上创造出第一台电子管计算机，是技术发明。后来创造出晶体管计算机，就算是技术革新，当然现在的集成电路计算机也算是技术革新，尽管其中的晶体管、集成电路本身是技术发明的成果。

技术革新虽然不同于技术发明，但与技术发明有着密切的联系。将技术发明成果应用于生产，是技术革新。而技术革新产生出新产品和新方法，又会促进新的技术发明。因此，技术革新和技术发明是技术领域里相互依存、相互促进的两种应用技术开发的典型创造活动。

科学发现、技术发明和技术革新是科技创新的三个活动内容。在总结、分析科技领域创新成果的基础上，人们提出了适合该领域的创新技法。

### 3.4.2 科技创新技法 Techniques of technological innovation

1.系统设问法

以系统提问方式打破思维的束缚，最大限度地发挥发散思维，扩展设计思路，提高人们创新性的设计能力。系统设问法是基于以下基本原理形成的：①世界上不存在不能加以改进的人工制品；②怀疑是进步的阶梯，怀疑已有产品的完善性是改进老产品、创新产品的前提；③提问可以开启人类智慧的闸门，引起人们的思考和想象，激发创新冲动，扩展思维。系统设问法的具体形式很多，下面介绍几种典型设问技法。

1)奥斯本设问法(奥斯本检核表法)。对不同产品的设计，创新内容是方方面面的，为了使思维系统化，美国创造学家奥斯本建议从不同角度设问，他归纳了九个方面，并列成一张目录表，逐条检索、设问。因此该方法又称奥斯本检核表法。

·转化：现有东西能否改做他用？是否有其他用途？

例如把电视机制成烘干机，蒸汽熨斗改成加湿器。

·引申:从现有的引申、移植出新的产品或技术。

例如医生把爆破技术用于治疗结石,将电子计算机引入机械产品,将圆珠笔引入到钢笔中。

·改变:改变原有的结构、材料、功能、形状、颜色、气味等,会有什么效果?

例如将滚动轴承中的滚动体变化,出现了滚子轴承、球轴承等;在饮料中改变配方出现了绿茶、红茶、橙汁等系列饮料。

·放大:能否放大?放大后性能如何?能否增加什么?高一些、大一些、长一些行吗?

例如两块玻璃之间加一层膜,可做成防碎玻璃;奔驰公司曾制造了13m长的轿车,其后部带一心形浴池,供行驶中使用。

·缩小:是否可以缩小?使之变小、浓缩、压缩、降低、变低、变轻、变薄、变短,会有什么结果?是否可以分割?

例如超浓缩洗衣粉、应用集成电路技术制成的微电子产品。

·颠倒:可以颠倒使用吗?正反、上下、左右颠倒后会有什么影响?

例如装配工序颠倒,制造电池工序颠倒变为制氢工艺。

·替代:是否可以对原有产品的材料成分、功能工序替代?

例如石英手表替代机械手表、全塑汽车概念、电动汽车概念。

·变换:改变模式、排列、布置形式或因果关系,速率、时间、材质等会有新的发现吗?

例如服装面料、花型、领子等稍做变换,可设计出新的款式来。

·组合:现有技术能否组合在一起?是整体还是部分组合?是功能、材料原理组合吗?

例如手机与照相技术组合成可拍照手机,杯子与茶叶桶组合成方便旅行杯。

以保温瓶为对象,运用奥斯本设问法提出新产品设想,结果如表3-2所示。

**表3-2 奥斯本检核表应用**

| 序号 | 检核项目 | 新产品名称 | 设想要点 |
|---|---|---|---|
| 1 | 有无其他用途 | 保健理疗瓶 | 利用保温瓶热气对人体进行理疗,可止痛、预防感冒 |
| 2 | 能否借用 | 电热式保温瓶 | 借用电热壶原理制成电加热保温瓶 |
| 3 | 能否改变 | 个性化热水瓶 | 按照个性要求设计外观满足不同消费者的心理 |
| 4 | 能否扩大 | 双层瓶盖保温瓶 | 扩大瓶盖,分为两层,上层放茶叶 |
| 5 | 能否缩小 | 新型保温杯 | 将保温瓶缩小成多种保温杯 |
| 6 | 能否代用 | 不锈钢胆保温瓶 | 用不锈钢做胆 |
| 7 | 能否重新调整 | 新潮保温瓶 | 调整尺寸比例,使之标新立异 |
| 8 | 能否颠倒 | 倒置式热水瓶 | 变瓶口朝下 |
| 9 | 能否组合 | 多功能保温瓶 | 将保温瓶与花瓶、负离子发生器组合成一体 |

在科学发现史中,有许许多多的例子表明创造者自己无意识地运用了奥斯本设问法。让我们通过下面几个例子来看看科学家是怎样运用设问法进行创造活动的。

例1:1900年,英国物理学家瑞利根据经典统计力学和电磁理论,推出了黑体辐射的能量分布公式。该理论在长波部分与实验比较符合,但在短波部分却出现了无穷值,而实验结果是趋于零。这部分严重的背离，被称之为“紫外灾难”。怎么办？能否重新调整？德国物理学家普朗克据此对公式进行了调整，得出了一个在长波和短波部分均与实际相吻合的公式。在此基础上，为了合理地解释公式，普朗克提出了“能量子”或“量子”的假说，于是宣告了量子论的诞生。有无其他用途？第一个意识到量子概念还有其他用途的是爱因斯坦，他建立了光量子论以解释光电效应中出现的新现象。真正意识到量子概念可以推广到一切物质粒子，特别是电子的是法国物理学家德布鲁意。他连续发表了三篇论文，提出了电子也是一种波理论，并预言电子束穿过小孔时也会发生衍射现象。1924 年，他写出博士论文“关于量子理论的研究”，更系统地阐述了物质波理论。不出几年，实验物理学家真的观测到了电子的衍射现象，证实了德布鲁意物质波的存在。仅此而已吗？是否可以修正？是否有什么新想法？是否可以补充？是否可以有其他变化？德国物理学家德拜提出了另一问题：如果电子是波，它将服从什么波动方程？1926 年，奥地利物理学家薛定谔得出了与实验证据非常吻合的波动方程。1925 年，德国青年物理学家海森伯创立了解决量子波动理论的矩阵方法，玻恩随后将其发展为矩阵力学理论。1926 年 3 月，薛定谔证明波动力学与矩阵力学在数学上是完全等价的。这样量子理论就有了两种表述方式。

例 2：19 世纪中叶，中国从西方引进了铅活字印刷技术，这种印刷方法一直主宰了中国印刷业 100 年，它不仅劳动强度大，排版的效率非常低，且铅的污染很大。1981 年，王选主持研制成功我国第一台计算机汉字激光照排系统原理性样机华光Ⅰ型。1982—1993 年，国家科技进步一等奖等荣誉开始接踵而至。面对这些荣誉，王选并没有停止思考，中国的印刷技术仅此而已么，还能不能修正？是否有新想法？他又先后主持研制成功并推出了华光Ⅲ型机、Ⅳ型机到方正 93 系统共五代产品，以及方正彩色出版系统。

1985 年，王选教授发明了用数学描述的方法对汉字进行分解压缩铸字储存，解决了把汉字输入计算机压缩还原技术，极大减少了汉字数据的存储量，使我国的计算机激光照排得以实现。把一张报纸排版所需的时间从以前的 4h 降到 20min，从此，我国的印刷技术迎来了计算机与激光的时代。王选的发明被称为中国印刷术的第二次革命。但是是否还可以有别的变化？1991—1994 年，王选领导北大方正不断创新又引发了报业和印刷业三次技术革新：告别报纸传真机，直接推广以页面描述语言为基础的远程传版新技术，致使我国报纸的质量和发行量大大提高；告别传统的电子分色机阶段，直接研制开放式彩色桌面出版系统，引起一场彩色出版技术革新；告别纸和笔，采用采编流程管理的电脑一体化解决方案。

例 3：17 世纪的一个夏天，英国著名化学家波义耳正急匆匆地向自己的试验室走去，刚要跨入实验室大门，阵阵醉人的香气扑鼻而来，他这才发现花圃里的玫瑰花开了。他本想好好欣赏一下迷人的花朵，但想到一天的实验安排，便摘下几朵紫罗兰插入一个盛水的烧瓶中，然后开始和助手们做实验。不巧的是，一个助手不慎把一滴盐酸溅到紫罗兰上，爱花的波义耳急忙把冒烟的紫罗兰用水清洗了一下，重新插入花瓶中。谁知当水落到花瓣上后，溅上盐酸的花瓣奇迹般地变红了，波义耳立即敏感地意识到紫罗兰中有一种成分遇盐酸会变红。那么，这种物质到底是什么？别的植物会不会有同样的物质？别的酸对这种物质会有什么样的反应？这对化学研究有什么样的意义？这一奇怪的现象以及一连串的问题，促使波义耳进行了许多试验。由此他发现，大部分花草受酸或碱的作用都会改变颜色，其中以石蕊地衣中提取的紫色浸液最为明显，它遇酸变成红色，遇碱变成蓝色。利用这一特点，波义耳制成了实验中常用的酸碱试纸——石蕊试纸。在以后的 300 多年间，这种试纸一直被广泛应用于化学实验中。

还可以提些什么问题？就留给读者来思考吧。运用检核表法，应该掌握几个要点：①要有条理地进行提问思考，检查一下是否毫无矛盾地列入所有的项目。②要以过去的事例、他人的意见以及其他的信息为基础来列出项目。③根据检核进行设想时，要时常反过来检查一下是否有不周之处。④最好是先制作成设想卡片，再进行排列组合整理。⑤可以只选择检核表中的1～2条，单独提问思考。

2)5W2H法。针对某个要解决的问题（产品、方法等），从以下角度系统发问：①Why：为什么要设计该产品？为什么要采用这种结构？从而明确目的、任务、性质。②What：该产品的功能是什么？有什么方法、手段设计该产品？③When：什么时候完成该产品的设计？设计各阶段时间的确定。④Where：该产品是针对哪个地区、哪个行业开发的？在哪里生产？在哪里鉴定？⑤Who：该产品终端用户是谁？谁来完成设计？由一个人来完成还是组成设计小组？⑥How to do：如何设计？产品的结构、材料如何取？颜色、造型风格如何定？⑦How much：是批量生产还是单件定做？产品价值是高还是低？

该方法实质上把奥斯本设问法浓缩为7条设问。

2.系统列举法

系统列举法是通过尽可能详尽列举待改进、发明、设计的产品的各种特性，以全面系统地解剖产品，把问题化整为零寻求突破的一种创新技法。

1)特性列举法。美国内布拉斯加大学Crawford教授认为创新实际上就是对旧事物某些特征的继承和改变。为了寻求需要改进的方向，首先应该全面列举事物的属性和特征，然后逐一分析，打通思路，得出创新方案。列举时，可从三个方面入手：

- 列名词特征：产品的组成部分、材料等。
- 列形容词特征：产品的性质、形状、色彩等。
- 列动词特征：产品的功能、作用、意义等。

例如要改良或开发一种新的自行车品种，从整体上着手比较困难，往往不易找到突破口，若采用特性列举法，则可以很容易得到一些新思路。

名词特征：①组成：车架、车轮、链传动机构、车胎等；②材料：钢、合金钢、塑料。

形容词特征：①颜色：黑、白、红、天蓝等；②结构形状：男车、女车、单梁、双梁，28″、26″、24″、22″，加重的、轻便的、变速的、手闸的、脚闸的等。

动词特征：①从功能看，有单人骑座、广告展示车、杂技表演车、观赏车、儿童玩具车；从附加功能看，可载货、带小孩、放物品；②从动力看，有人力、电力、风力等。

如此层层分析，区分出事物的属性之后，再针对每一种属性运用检核表法提出新设想就容易得多了。

2)缺点列举法。没有绝对完美的产品，社会总在发展进步，人们对生活的追求是不断提高的，当发现了现有事物、设计等的缺点，找出改进方案，便可设计出新的产品。

例如汽车公司针对冬天坐垫冰冷，开发出了可预热的热垫椅；针对驾驶员长时间开车会引起疲劳，开发出具有按摩功能的座椅；针对驾驶员操作方向盘总是一个固定坐姿，开发出了可调角度的方向盘。针对汽车超载行驶，有人提出超载时使汽车报警；又如残疾人用的轮椅不但要在平地行走，还要考虑上楼梯、过天桥，因此人们开发设计出了一种履带式轮椅，可兼顾上述两种功能。

3)希望点列举法。不同于缺点列举法，希望点列举法不是对现有物品进行改进的被动方法，而是从发明者、用户的意愿出发提出新设想的积极、主动型的方法。

例如人们希望打电话时不仅可闻其声，而且能见其面，从而开发了可视电话。例如手机目前属于一个普及性的通信工具，虽然通信是其基本功能，但为了满足人们追求时尚、个性的心理，厂家已开发出商务手机、可拍照手机等。

运用希望点列举法，首先要了解消费需求的客观趋势与特点。人们将这种趋势与特点归纳为9个方面：追求舒适的生活；追求美的倾向；追求文化教育的倾向；讲究格调的倾向；希望实惠的心理；追求时尚、流行的心理；喜好美食的倾向；重视健康的心理；追求知识的心理。

3. 形态分析法

按材料分解、工艺分解、功能分解、形态分解、成本组成分解，把待解决问题分解成各个独立的要素，然后用图解法将要素进行排列组合，从许多方案的组合中找出最优解，便是形态分析法，又称“形态矩阵法”“形态综合法”。天文学家茨维克在参与美国火箭研制过程中，用形态分析法，按火箭各主要组成部件可能具有的各种组合，得出上千种火箭构造方案，其中不少极有价值，并在方案中包含了当时德国正在研制而严加保密的带脉冲发动机的F－1型巡航导弹和F－2型火箭。此方法的实际操作如下：

首先明确要解决的问题，其次列出要解决问题的有关独立因素，再次详细列出各独立因素所包含的要素，最后将各要素排列组合成创新设想。

例如设计制造一种运输物品的新型工具形态矩阵明细如表3－3所示。

**表3－3　运输物品的新型工具形态矩阵**

| 装载形式 | 输送方式 | 动力来源 |
|---|---|---|
| 1. 车辆式 | 1. 水 | 1. 压缩空气 |
| 2. 输送式 | 2. 油 | 2. 蒸汽 |
| 3. 容器式 | 3. 空气 | 3. 电动机 |
| 4. 吊包式 | 4. 轨道 | 4. 风力 |
| 5. 其他 | 5. 滚轴 | 5. 蓄电池 |
|  | 6. 滑面 | 6. 内燃机 |
|  | 7. 管道 | 7. 太阳能 |
|  | 8. 其他 | 8. 其他 |

此例可组合5×8×8＝320个方案，从中筛选可行的方案再细化。

4. 类比创新技法

世界上的万物千差万别，但并非杂乱无章，它们之间存在程度不同的对应关系：有的是本质类似，有的是构造类似，也有的仅有形态、表面的类似。从异中求同，从同中见异即可得出创新成果。

类比是通过比较事物之间在局部上的相同或相似而把性质上不同的事物联系起来的一种思维方法，虽然其结论带有较大的偶然性，但能给人以启发。类比可以在严格的意义上运用，如类比推理，其基本要求是尽可能多地寻找事物之间的相同点或相似点，以便类比推理，它有多种形式，如形象类比、关系类比、结构类比、功能类比等。类比也可以在不太严格的意义上使用，如比喻、比拟等。和严格意义上的类比相比较，比喻、比拟更富有诗意，也更灵活、更自由，并且它们的重点不在于相似性，而在于当两个关系不同的事物彼此都具有某种品质时所产生的激情。这有助于将表面上看不相关的东西联系起来，从而使大脑有更多的机会去选择导致新的顿悟联想。

人们用得最多且行之有效的就是直接类比，它的典型方式是功能模拟和仿生。从这个意义出发，可对类比创新技法进行分类，如图 3－3 所示。

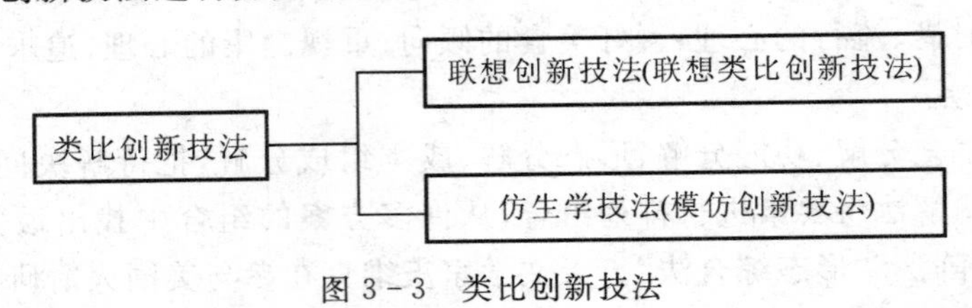


图 3－3　类比创新技法

联想创新技法，就是借助各种联想思维形式产生新的设想，得到创新成果。

大自然造就的千万种生物是它们同周围环境斗争中，通过优胜劣汰而长期进化形成的。研究生物的习性、结构、行为可以给技术工作者以很大的启发。20 世纪 60 年代以来出现的仿生学便是这样的一门新学科。仿生学的具体定义为：为解决技术上的难题而应用生物系统的知识的学问。

运用仿生学的思想进行创新的方法，就是仿生学技法。

日本发明家田熊常吉将人体血液循环系统中动脉和静脉的不同功能和心脏瓣膜阻止血液逆流的功能运用到锅炉的水和蒸汽的循环中，发明了田熊式锅炉，热效率提高了 10%。法国艺术家莫尼哀，长期观察、研究植物，培育出了许多优良品种。他在园艺工作中，看到水泥制成的花坛经常被碰碎，想找个办法解决这个问题。他发现植物盘根错节的根系能够使松软的泥土变得坚固，就产生了联想：用铁丝模仿植物的根系，水泥模仿土壤，让铁丝交叉成网状再用水泥将铁丝包起来，制成的花坛一定会很坚固，实践的结果证明莫尼哀制作的花坛结实异常，这就是最早的钢筋混凝土结构。

有一个自古相传的故事，说的是我国的木匠祖师鲁班。他一直在为木材的分解伤脑筋，用什么样的工具分解木材呢？一天，他上山砍柴，不知怎么搞的，手指被茅草割开了一个小口子，鲜血从小口中渗了出来。他抑制不住好奇心，仔细观察起来：为什么看起来很柔软的茅草，能割破手指？突然，他看到茅草的边缘是波浪起伏的齿状。或许是这些齿的作用吧？他又用茅草在手指上试了一下，对了，就是这个缘故，如果将铁片的边缘也做成这样的齿状，它是否就可以割开木材呢？灵光闪耀之后，赶紧回去试验，成功了。鲁班因此发明了锯子。

人们发现响尾蛇的鼻和眼的凹部中具有敏感的检测出红外线的能力，其表面能对 1‰℃的温度变化做出反应。因此，响尾蛇能轻而易举地觉察到身边其他动物的存在。就人的皮肤而言，即使是对温度最敏感的部分，至多也只能对 1/10℃的温度变化做出反应，美国的菲尔科公司和通用电气公司根据这一原理，研制出了“响尾蛇”导弹。

面粉中加入发酵粉能使体积增大，人们在塑料中加入发泡剂，生产出了省料轻质的泡沫塑料。从对人类本身的拟人类比，亦会产生许多设计、创造，如设计各种机器人，有的能自动喷漆、自动焊接，有的机器人会下棋，帮助人做家务等等。

### 5.头脑风暴法

这是美国创造学家奥斯本于1901年提出的最早的创新技法，又称智力激励法、激智法等，它是一种群体集智法。一般是通过一种特殊的小型会议，使与会人员围绕某一课题相互启发、激励，取长补短，引起创造性设想的连锁反应，产生众多的创造性成果。

该技法的模式如下：

· 形式：5～10人的畅谈会。

· 时间：30～45min，最长不超过1h。

· 议题：单一问题，对于复杂问题可分解为若干子问题，多次会议解决。

· 原则：a. 自由奔放，打破定势；

b. 不许评判和质疑；

c. 追求数量，再从数量中求质量；

d. 借题发挥，相互启发。

· 记录、整理、会后评价。

具体实施该技法还应注意以下问题：

1)选择好会议主持人。智力激励会议的成功或失败在很大程度上取决于主持人掌握会议的方法。因此，要求所选的主持人具有较丰富的智力激励经验，能把握住会议的主题。主持人掌握会议时要把握住三点：

· 严格遵循智力激励会议的四条基本规则。

· 要使会议保持热烈的气氛。

· 要让所有会议参加者都能献计献策。

2)选择好会议的记录员。会议的记录员必须及时记下大家提出的新设想，并写在醒目的位置上，让大家都能看到。同时，记录员还要把设想记录在小卡片上，编号保留下来。一个记录员不够时，可以配两名记录员。

3)确定好会议的主题。主持人应在会议召开的前两天将主题通知参加者，并附上必要的说明，让参加者能够收集确切的资料，按照正确的方向思考问题。

4)选择好会议的参加者。

· 奥斯本认为会议的规模宜控制在5～10人，日本一些创造学家认为6～9人合适，我国的创造学家认为5～15人合适。但要注意，太多则容易产生分歧，太少则涉及面会过窄。

· 参加者的专业结构不能单一化，要尽量使专业结构多样化。

· 要适当选择有实践经验的参加者，以便形成核心小组。

5)组织好会议。会议开始前，为使参加者尽快将精力转移到会议上来，可以举行热身活动。其内容有多种形式，如让参加者看一段有关创造的录像，讲一个创造性强的小故事，做几道“脑筋急转弯”的思维练习题。

会议开始时，主持人应简明扼要并富有启发性地向参加者介绍要解决的问题，然后让参加者自由畅谈，提出富有创造性的新设想。此时，主持人要牢记自己的职责。

### 6.对智力激励法的改进

人们在运用智力激励法进行创造的过程中，根据自己的情况对该技法进行了改进。这里我们介绍几个典型的改进技法。

1)635法。该技法又称为默写式智力激励法，是德国创造学家霍利格根据德意志民族惯于沉思、不喜高谈阔论的性格特点提出的。其要点是将参加会议的人数和会议的自由畅谈过程改变为：

·参加者以 6 人为最理想。

·每人在设想卡上写 3 个设想。

·写 3 个设想的时间是 5min。

·5min 一到，就将设想卡交换一下。这时与会者围成一圈，交换卡可按一定的方向（如逆时针方向）进行。

·再重复上面 3 个步骤，经过 30min 达到一个循环，可得到 108 个设想。

·将卡片收齐，分类整理再进行评价，最后选择出有价值的设想。

2)MBS 法。日本的三菱树脂公司在运用智力激励法时，提出了另外一种改进方案。其实施过程是：

·主持人向参加会议的人宣布主题。

·给 10min 左右时间，让参加者将设想写在笔记本上。

·轮流宣读设想，每人每次宣读 1～5 个设想，记录员记下设想，其他人受到启发后又可在笔记本上写下新设想，尽量让全体人员把所有设想宣读完。

·开始对设想提出质询，提设想人进行说明。

·主持人对讨论结果进行归纳。

·参加者对设想进行评价，整理出有用的设想来。

3)CBS 法。它是日本创造开发研究所所长高桥诚提出的一种改进方案。其实施过程为：

·主持人宣布主题。

·参加者（6～8 人）围坐在桌子周围，每人拿 50 张左右卡片写设想，每张卡片写一个设想（约占 1/6 的时间）。

·轮流宣读卡片，并排列在桌面上，读一张放一张，别人可以提出质询，并在受到启发时将设想写在自己的卡片上（约占 1/2 的时间）。

·自由发表设想，从桌上拿掉有重复的设想卡（约占 1/3 的时间）。

·主持人进行归纳。

4)KJ 法。这是日本职工创造活动中选用位居第一位（75%）的创造技法，是日本东京工业大学教授川喜田二郎于 1964 年提出的对智力激励法的一种改进方案。其实施过程如图 3-4 所示。

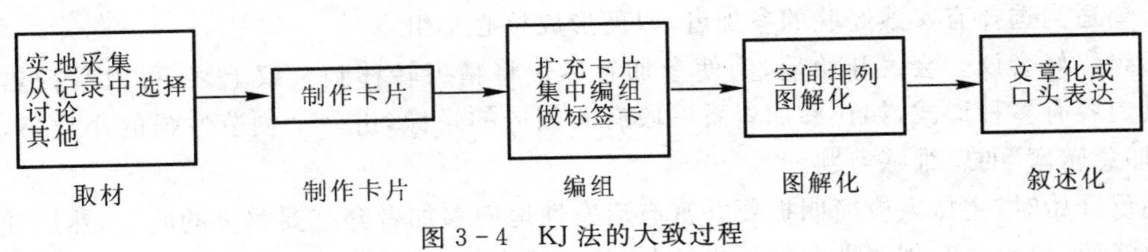


图 3-4　KJ 法的大致过程

·准备工作：主持人一名，与会者 4～8 人，准备好卡片和黑板。

·获取设想：按奥斯本智力激励法进行，以获取 30～50 条信息或设想为宜。

·制作卡片：将搜集到的信息或设想编成两行左右的短语，写到卡片上，每张卡片上写一条信息或设想。这样制作的卡片叫做“基础卡片”，每个与会者抄录一套。

·整理卡片:这时要分成三步来进行:

第一步,分成小组。由与会者按照自己的思路各自进行卡片分级,把内容在某点上相同的卡片归在一起,并给一个适当的合乎卡片内容的标题,写在一张卡片上,称为“小组标题卡”。不能归类的,每卡自成一组。

第二步,并成中组。将每个人所写的小组标题和自成一组的单卡都放在一起,让与会者共同讨论,将内容相近的小组卡片归在一起,再给一个适当的标题,写在一张卡片上,称为“中组标题卡”,不能归类的自成一组。

第三步,归成大组。经共同讨论,再将中组卡和单卡放在一起并归成大组,再给一个适当的标题,从而获得“大组标题卡”。

·综合求解:将整理出来的卡片,以其隶属关系,固定于黑板或贴在纸上,并用线条将有关项目固联,即可形成综合方案的图解,然后按图解形成文字,以表述比较完整的新设想方案。

仔细研究,可以看出 KJ 法是将智力激励法对设想整理评价工作完善后的结果。对智力激励法的改进,还有许多种方案,限于篇幅,就不一一介绍了。

*7.综摄法*

该方法又称分合法、集思法,其基本操作是:在一位主持人召集下,由数人构成一个小组,小组成员的专业不同、知识背景不同。会议主题往往是把具体问题提炼后的抽象、简单词汇,运用隐喻、类比可操作性的方法,调动小组成员的潜意识,把不熟悉的东西变为熟悉的东西,或把熟悉的东西变为不熟悉的东西,并在相互启发下提出创新的思路。

例如火车站附近开发一个自行车停车场。主持人一开始仅提出一个词汇——存放。小组成员就存放提出设想:“放进竹筒去”“存到银行去”……主持人点破主题——开发停自行车的车场。小组成员根据上面的设想,围绕主题可得出许多方案。如“放进竹筒里去”的设想,可启发为:车站附近建塔式建筑存放;月台下空地洞存放。“存到银行里去”的设想,可启发为:在车站附近建存车处,按取车的时间先后分别归类,用卡车先将自行车运到别处空地上,到时候再运回来交存车人等。最后对各方案检验评价、细化,得出最佳方案。

*8.六顶思考帽法*

运用水平思考法,仅允许思考者在同一时间内只做一件事情,强迫思考者将逻辑与情感、创造与信息等区分考虑,戴不同帽子使人们依次对问题的不同侧面给予足够充分的考虑。最终得出事物的全方位“彩色”答案,每一顶帽子的颜色与其功能是相关的。

·白色思考帽:白色代表中性和客观。白色思考帽思考的是客观的事实和数据。

·红色思考帽:红色代表情绪、直觉和感情。红色思考帽提供的是感性的看法。

·黑色思考帽:黑色代表冷静和严肃。黑色思考帽意味着小心和谨慎,它指出了任一观点的风险所在。

·黄色思考帽:黄色代表阳光和价值。黄色思考帽是乐观、充满希望的积极的思考。

·绿色思考帽:绿色是草地和蔬菜的颜色,代表丰富、肥沃和生机。绿色思考帽指向的是创造性和新观点。

·蓝色思考帽:蓝色是冷色,也是高高在上的天空的颜色。蓝色思考帽是对思考过程和其他思考帽的控制和组织。

水平思考法是英国生态生理学家博诺在其著作《水平思考的世界》中介绍的技法。它的运用要点是:

1)要找到支配的观点。所谓“支配性的观点”是指推动着现实生活的现成的概念。虽然现成概念未必采取明确的形态,但是,对于创造活动来说,它却是很大的障碍。为了产生新的设想,首先就要消除这些障碍。因此要有意识地找出支配现状的观点,并使之明确化,然后指出其弱点。这样一来,就可以避开现成的概念,自由地进行设想。

2)寻求各种观点。一般而言,处理问题时,首先要规定解决问题的范围,只有在这个范围内,才能采用累积逻辑的方式。但是,实际的解决办法在该范围之外的情况也很多。例如哥伦布发现美洲大陆后,一些人不以为然,认为这是任何人都可以办到的,并对哥伦布进行讽刺、刁难。为了教育这些人,哥伦布拿出一个鸡蛋对这些人说:“你们谁能把鸡蛋竖起来?”这些人全都束手无策。于是,哥伦布敲破鸡蛋后,把它竖了起来。并对这些人说:“当我做过之后,你们任何一个人也就都会做了。”如果哥伦布被不可敲破鸡蛋这一条件所束缚,那问题就无法得到解决。

寻求各种观点,可以这样去做:

·预先决定事物因果关系的数量,这个数量以3～5个为宜。决定数量以后,碰到问题时就可以有意识地在3～5个因果关系中找到一个相应的因果关系。

·有意识地推翻事物的因果关系。例如,可以这样考虑,即房屋的支柱不是用来支撑屋顶的,而是从屋顶上悬吊下来的。

·使状况变得更容易处理,这样就可以消除最初状态中的限制,使自由设想更容易进行。在这一过程中产生的新设想可与原有状况对比,看看是否适应。

·把侧重于一个问题的某个部分的视点移到其他部分上去。如果改变问题的着重点或集中点的话,那么观点就会自然而然地改变,并且能够产生新的观点,解决的办法也就要变化。

3)摆脱垂直思考模式的束缚。当萌发新设想时,垂直思考不仅毫无作用,而且还会抑制设想的产生。这是因为最初忽然闪现的设想并不是俯拾即是的,虽然在逻辑上难以说明,但如果在这种场合使用垂直思考的方法,无疑会扼杀好不容易萌发出来的设想。

4)灵活地利用偶然的机会。博诺认为,赫兹发现无线电波,伦琴发现X射线,塔贝尔发明使用氯化银的照相感光板等等都是他们灵活地利用偶然性因素的结果。

那么,如何灵活地利用偶然性呢?如果习惯于水平思考,就可以利用偶然获得信息,并能逐渐擅长于萌发偶然情况下的有关设想。这也就是要时常考虑那些与要害问题无关的其他因素,即使根据仅有的一点信息,也常常可以使设想得到飞跃性的发展。

种种技法,是前人经验总结的、实践证明是行之有效的。掌握与运用这些技法,是取得创新成果的手段和捷径。然而要成为一个成功的创新设计师必须具备较强的觉察能力、记忆能力、联想能力、分析与综合能力、想象能力、自控与完成能力。因此对工业设计师来说,必须通过不断学习、训练、实践,逐渐培养自己的创新能力。

## 3.5 创新法则 Innovation rule

上一节按类别,按思维习惯不同介绍了一些常用创新技法,此外,还有如灵感法、废物利用法、专利利用法、移植法、组合法、功能分析法等,许多方法具有内在的相似性。在创造过程中,不要机械地使用某种技法,根据要求不同,一种技法可以重复使用数次,也可以同时使用几种技法。通过学习前面的创新技法,我们不难发现创新和创造是有其基本规律和法则的。可将其归纳如下：

1.综合法则

这是指在分析各个构成要素的基础上加以综合，使综合后的整体作用促成创造性的新成果。这种综合法则在设计、创新中广为应用。它可以是新技术与传统技术的综合，可以是自然科学与社会科学的综合，还可以是多学科成果的综合。如计算机，即综合了数学、计算技术、机电、大规模集成电路技术等方面的成果。人机工程学是技术科学、心理学、生理学、社会学、卫生学、解剖学、信息论、医学、环境保护学、管理科学、色彩学、生物物理学、劳动科学等学科的综合。美国的"阿波罗"登月计划可算是当代最大型的各种创造发明、科学技术的综合，该项计划准备了 10 年，动员了全美国 1/3 的科学家参加，2 万多个工厂承做了 700 多万个零件，耗资达 240 亿美元。

2.还原法则

人的头脑并无多少优劣之分，起决定作用的是抓取要点的能力和丢弃无关大局的事物的胆识。还原法则即抓事物的本质，回到根本，抓住关键，将最主要的功能抽出来，集中研究其实现的手段与方法，以得到有创造性的最佳成果。还原法则又称为抽象法则。

洗衣机的创新成功，是还原法则应用的成功例子。其本质是"洗"，即还原。而衣物脏的原因是灰尘、油污、汗渍等的吸附与渗透。因此，洗净的关键是"分离"。这样，可广泛地考虑各种各样的分离方法，如机械法、物理法、化学法等。根据不同的分离方法，因而创造出了不同的洗衣机。我们不妨设想一下，为什么汽车一定是四个轮子加一个车身呢？为什么火车一定是车头拉着车厢在铁轨上滚动呢？因为交通运输工具的本质，应该是将人、货从一处运到另一处。同样是火车，却可以是蒸汽机车、内燃机车、电力机车或是磁悬浮列车。还原到了事物的创造起点，相信会有与现今不同形式的交通运输工具设计创造出来。

3.对应法则

俗话说"举一反三""触类旁通"。在设计创造中，相似原则、仿形仿生设计、模拟比较、类比联想等对应法则用得很广。机械手是人手取物的模拟；木梳是人手梳头的仿形；夜视装置即猫头鹰眼的仿生设计；船和潜艇来自人们对鱼类和海豚的模仿。火箭升空利用的是水母、墨鱼反冲原理。通过研究变色龙的变色本领，为部队研制出了不少军事伪装装备；用两栖动物类比得到水陆两用工具……这些事例均属对应法则。

4.移植法则

这是把一个研究对象的概念、原理、方法等运用于另一研究对象并取得成果的有效法则。"他山之石，可以攻玉。"应用移植法则，打破了"隔行如隔山"的禁界，可促进事物间的渗透、交叉、综合。

日本开始生产聚丙烯材料时，聚丙烯薄膜袋销路不畅。一天，推销员吉川退助在神田一酒店休息时，女店主送上手巾给他擦汗，因是用过的毛巾，气味令他厌恶。他突然想到：如果每块洗净的湿毛巾都用聚丙烯袋装好，一则毛巾不会干掉，二来用过与否一目了然。于是申请了小发明，仅花了 1 500 日元，而获利高达 7 000 万日元。

上海原有 104 万只煤饼炉，居民为晚上封炉子而烦恼。封得太紧，白天起来已灭掉；封得稍松，早上煤饼已烧光。一位中学生，将双金属片技术移植到炉封上，发明了节能自控炉封，使封口间隙随炉内温度而自动调节，既保护了封炉效果，也大大节省了煤饼。

蚕丝是中国的特产，但目前面临着日本的挑战。日本人培养出新型的蚕种不吃桑叶改吃苹果、胡萝卜，创造了奇迹，被称之为"日本蚕"，使之成为日本的特色，这就是日本人的聪明。

移植的方法亦可有所不同，可以是沿着不同物质层次的"纵向移植"，也可以在同一物质层次内不同形态的"横向移植"，还可以多种物质层次的概念、原理、方法综合引入同一创新领域中的"综合移植"等。

5．离散法则

上述的综合法则可以创新，而其矛盾的对立面——离散——亦可创新。这一法则即冲破原先事物面貌的限制，将研究对象予以分离，创造出新概念、新产品。隐形眼镜即眼镜架与镜片离散后的新产品。音箱是扬声器与收录机整体的离散；活字印刷术即是原来整体刻板的分离。为了节约木材，将火柴头与火柴梗分离，在火柴头内加铸铁粉，用磁铁吸住一擦就燃，应用了离散法则，即发明了磁性火柴。

6．强化法则

强化法则又称聚焦原理。如利用激光装置及专用字体创造的缩微技术，使列宁图书馆20km长书架上的图书，缩纳在10个卡片盒内。对松花蛋进行强化实验，加入菊花、山楂及锌、铜、铁、碘、硒等微量元素，制成了食疗降压保健皮蛋。两次净化矿化饮水器，采用了先进的超滤法，含有5种天然矿化物层，大大增强了净化矿化效果，还能自动分离、排放细菌及污染物。仅用一滴血在几分钟内就可做10多项血液化验的仪器、浓缩药丸、超浓缩洗衣粉、增强塑料、钢化玻璃、采用金属表面喷涂或渗碳技术以提高金属表面强度等，均是强化法则的应用。

7．换元法则

换元法则即替换、代替的法则。在数学中常用此法则，如直角坐标与极坐标的互换及还原、换元积分法等。达维道夫用树脂代替水泥，发明了耐酸、耐碱的聚合物混凝土。亚·贝尔用电流强度大小的变化代替、模拟声波的变化，实现了用电传送语言的设想，发明了电话。高能粒子运动轨迹的测量仪器——液态气泡室——的发明，是美国核物理学家格拉肖在喝啤酒时产生的创造性构想。他不小心将鸡骨落到了啤酒中，随着鸡骨沉落，周围不断冒出啤酒的气泡，因而显示了鸡骨的运动轨迹。他用液态氢介质"置换"啤酒，用高能粒子"置换"鸡骨，创造了带电高能粒子穿过液态氢介质时同样出现气泡，从而能清晰地呈现出粒子飞行轨迹的液态气泡室，并因此获得了1979年的诺贝尔物理学奖。

8．组合法则

组合法则又称系统法则、排列法则，它是将两种或两种以上的学说、技术、产品的一部分或全部进行适当结合，形成新原理、新技术、新产品的创造法则。这可以是自然组合，亦可是人工组合。

同是碳原子，以不同处理、不同晶格的结合，便可合成性能、用途完全不同的物质，如坚硬而昂贵的金刚石和脆弱的良导体石墨。计算器用太阳能电池，装上日历、钟表，组合得到了新产品。不同金属与金属或非金属可组合成性能良好的各种复合材料。在煤饼炉炉底加上一导电加热的铁板，设计成了电热煤饼炉新产品，使引燃煤饼时不用木柴、纸张，也消除了滚滚浓烟。现代科技的航天飞机，即火箭与飞机的组合。20世纪80年代上海建筑艺术"十大明星"的"龙柏"饭店，因它在虹桥机场邻近，故建筑高度受限制。设计师在六层客房前用三层层高布置两层高的贵宾用房，使贵宾用房的室内空间更为舒适，贵宾休息室又设计成上面向内倾斜、呈1/4圆的台体，再加上波形瓦饰面的陡直屋面。这种高低组合、曲直几何组合的创新设计，使饭店具有新时代的特征而又不失民族特色。

组合创造是无穷的，但方法不外乎主体添加法、异类组合法、同物组合法及重组等四种。主体添加法就是在原有思想、原理、产品结构、功能等之中，补充新的内容。两种或两种以上不同领域的思想、原理、技术的组合，为异类组合法，这种方法创造性较强，有较大的整体变化。同物组合则在保持事物原有功能、意义的前提下，补足功能、意义，产生新的事物。将研究对象在不同层次上分解，以新的意图重新组合，称为重组，能更有效地挖掘和发挥现有科学技术的潜力。

9.逆反法则

一般来说，如果仅仅照人们习惯使用的顺理成章的思维方式，是很难有所创造的，因为就创造的本质而言，本身就是对已有事物的“出格”。应用逆反法则，即打破习惯的思维方式，对已有的理论、科学技术、产品设计等持怀疑态度，“反其道而行之”，往往会得到极妙的设计、创造发明。花园、环境绿化，顺理成章是在地面上。但应用逆反法则，现在下沉式、空中式、内庭式、立体绿化等比比皆是，由此创造了一种美好的生活空间。如果只想到“水往低处流”，就发现不出虹吸原理。别人都在炼纯锗，而日本的江崎于奈和宫原百合子却在锗中加杂质，产生了优异的电晶体而分别荣获“诺贝尔奖”。以0.1mm的药流在15MPa下注入皮下组织而毫无痛感的无针式注射器；人倒退走路，使脊柱相反受力而治疗腰肌劳损等疾病，亦是逆反法则在医疗技术上的应用。过去总说“生命在于运动”，而现在“生命在于静止”的静默疗法，让病人运用想象力来表达自己与疾病作斗争的愿望，用静功来运行全身气血，可使精神放松、改变体内生理生化状态、增强机体免疫功能，以战胜疾病。美国科学家发明了一种放在眼球上的长效眼药，可按控制的速度均匀地释放药效400h，以治疗青光眼等长期眼疾，一改了以往的供药方式。

在服装设计中，过去袖子、领子、口袋总是左右对称，如果要绣花、挑花，也是对称为主。而现在，袖子不同色彩，口袋左右不同，领子两面不一的服饰更显时尚。衣料亦一反常态，用水洗皱或用石子磨旧，也别具风格。

苏格兰一家图书馆要搬迁，图书馆发出了取消借书数量限制的通告，在短期内大量图书外借，至还书时还到新址，完成了大部分图书的搬运任务，节约了费用。这也是逆反法则、离散法则的实际应用。

一架飞机坏了，需要拖到机库里去修理，而拖是较麻烦的。于是，国外有人设计了一种“移动式机库”，可以开到飞机旁边进行修理，十分方便。将“飞机”的视点迁移到“机库”，使“飞机入机库”成为“机库纳飞机”。这些设计者真是聪明极了。

在一次香港小姐的决赛中，为了测试参赛小姐的思维速度和应对技巧，主持人提出了这样一个难题，“假如你必须在肖邦和希特勒两个人中间，选择一个作为终身伴侣的话，你会选择哪一个呢?”其中有一位参赛小姐是这样回答的：“我会选择希特勒。如果我嫁给希特勒的话，相信我能够感化他，那么第二次世界大战就不会发生了，也不会有那么多的人家破人亡了。”这位小姐巧妙的回答赢得了人们的掌声。这个问题难度较大，如果回答“选择肖邦”，则答案没有特色，显得俗气；如果回答“选择希特勒”，则很难给予合理的解释。那位小姐的精彩之处就在于既选择了出人意料的答案，又找出了合理而又充满正义感的理由。

上述这些事例都是逆反原则的实际应用，从表面看似乎不可理喻，最终却出乎人们的意料，能取得更好的结果，因此它常常给人一种不可思议的神奇感觉。

10.群体法则

科学的发展,使创造发明越来越需要发挥群体智慧,集思广益,取长补短。现代设计法也摆脱了过去狭隘的专业范围,需要大量的信息,需要多学科的交叉渗透。工业设计亦逐步从艺术家个体劳动的圈子中解放出来,成为发挥“集体大脑”作用的系统性的协同设计。因此,群体法则在设计、创造中越显其重要性。

据美国著名学者朱克曼统计,1901 年至 1972 年共 286 位科学家荣获诺贝尔奖。其中 185 人是与别人合作研究成功的,占人数的 2/3。而且,随着时间的推移,发挥群体作用的比例明显增加,在诺贝尔奖设立后头 25 年,合作研究而获奖者占 41%,第二个 25 年中,占 65%,第三个 25 年中则上升为 79%;控制论的创立者维纳,常用“午餐会”的形式,从各人海阔天空的交谈、思维中捕捉思想的新闪光点,激发自己的创造性。美国纽约布朗克斯高级理科中学单在 1950 年中,就出了 8 位蜚声世界的物理学博士,其中格拉肖、温伯格于 1979 年共获诺贝尔物理奖。这就是一种“共振”“受激”的群体效应。

这种创新包括两种形式:一种是将已有的创新技法运用于新的专业领域和新的创新问题;另一种是根据创新问题的需要,创造新的创新技法。

创新是工业设计的本质,开发创造力就是为了能够创新,随处随时留心创新,这样才会最终取得丰硕的创新成果。

# 第 4 章　产品设计实践
# Product Design Practice

## 4.1　色彩　Color

### 4.1.1　目的

1)使学生对“色彩”课程所涉及的有关概念有进一步的理解和掌握；
2)培养学生对调色基数的确定和实际动手能力；
3)使学生了解调色机及喷枪的特点和使用方法。

### 4.1.2　要求

1)要求学生确定所需调色基数并用调色机手动调出相应颜色；
2)在老师的引导下，使学生完成对色彩的选择、手动调制及表现效果的学习；
3)要求学生了解调色机与喷枪的特点与应用。

### 4.1.3　设备

手动调色机、混油机、喷枪、空压泵等。

### 4.1.4　掌握调色及喷色的流程

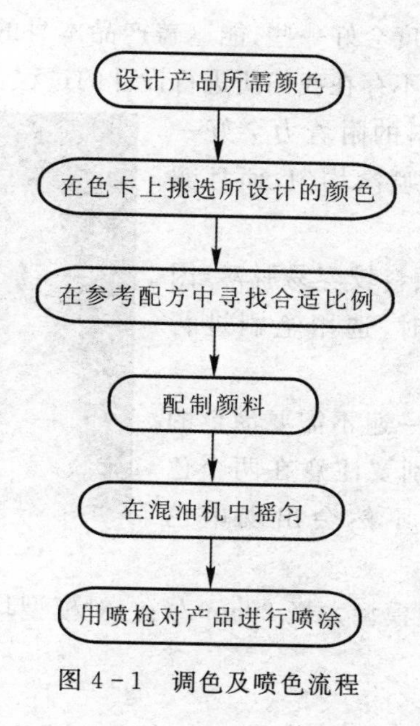

图 4-1　调色及喷色流程

### 4.1.5 具体步骤及注意事项

1)设计产品需要的颜色。

注意:在产品进行涂色之前,所做的模型必须进行表面处理。对模型进行涂色,则会将模型的表面处理进行一个放大的处理效果。表面光滑,则最后效果很好;如果有一点凹痕或者瑕疵之类的,则会将这些凹痕或者瑕疵效果进行放大,因此在涂色之前,一定要细心地进行表面打磨处理。

2)在色卡上挑选设计的颜色。

在色卡上挑选设计的颜色,例如:选择色卡上的翠绿色,色号为1172,GB10G 6/7.2。

3)根据色卡上的色号在参考配方中寻找合适比例。

在调色参考配方中先找到色卡编号为GB—1172的行,其中行中所示见表4-1。

**表4-1 色卡参考配方**

| 色卡编号 | 色浆1名称 | 色浆1用量/g | 色浆2名称 | 色浆2用量/g | 基钛白含量/(%) |
|---|---|---|---|---|---|
| GB—1172 | 1606钛青绿 | 1.06 | 1906中黄 | 0.1 | 18% |

就是取18%的乳胶漆,在其中加入1606钛青绿1.06g,加入1906中黄0.1g,进行混合,就可以得出GB—1172我们所设计出的翠绿色。

4)将所需各个颜色的颜料按比例取出,并在混油机中进行混合。

将各个颜色的颜料按比例取出,放在容器中混合,然后放在混油机中,操作设备,令其混合均匀。如果是少量的颜料进行混合,则用小棍搅拌均匀即可。

5)用喷枪对产品进行喷涂。

一般在对产品进行涂色过程中,有很多涂色方式。比如用自动喷漆、用喷枪进行喷漆、用笔刷等多种方式。

用自动喷漆进行喷涂,光泽度会好一些,能遮盖产品本身的质地与色泽。但是自动喷漆,只能买现成别人配好的颜色,而不存在我们所进行设计的过程。

用喷枪进行喷涂,这样颜料的附着力会好一些,如果喷涂在模型上面,不仅喷涂均匀,而且颜料不易脱落。

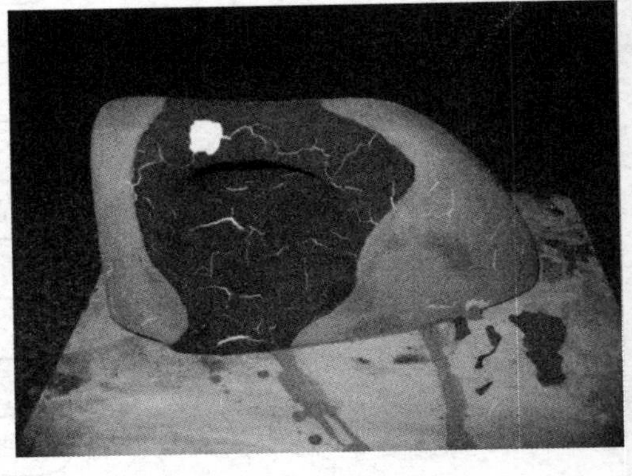

图4-2 色料脱落

用笔刷进行喷涂的时候,颜料较容易脱落,因此一般在模型的小范围涂色时,选用笔刷进行涂色。

在涂色之前,要注意,将第一遍不需要涂色的部位用胶带遮盖起来,而且特别要注意在两个色块交接的部位,一定要粘好,不然会出现串色现象。

喷涂应该为3遍,将喷头离模型大概20cm左右,对模型均匀地喷涂一遍,在喷涂的过程中，要注意把模型放在背风、人不经常走动的地方，而且模型的放置要保证在喷涂的过程中可以保证喷涂均匀。过10min左右，等第一遍的涂色基本吸收风干，进行第二遍喷涂，喷涂方式与第一遍相同，相同的程序进行3次，这样可以保证喷涂均匀，而且能完全掩盖住模型本来的质地与色泽。

#### 4.1.6　作业

1)将调色、喷色的程序详述一遍。

2)叙述用喷枪进行涂色和用笔刷进行涂色两种方式的涂色效果的不同点。

## 4.2　工业设计(一)　Industry design（Ⅰ）

### 4.2.1　项目一　产品认知

#### 4.2.1.1　目的及要求

1)通过实验，学生能够对“产品设计(一)”课程所涉及的有关设计方法有进一步的理解和掌握，根据认知产品创新特点检讨所用的创新设计方法。

2)使学生感受创新产品精巧之处，了解创新产品的巧妙构思。

3)通过实验，学生能够了解机电产品造型的特点。

4)通过实验，学生能够熟悉机电产品造型设计过程。

5)启发学生的创新意识。

#### 4.2.1.2　创新设计方法

创新方法是创造性思维的具体化。

创新本身就是探索性的社会实践活动，它有规律性，但带有模糊性。人们在模糊的规律指导下，尝试用创新设计方法解决问题。

常用的设计方法如下：

1.系统列举法

系统列举法是通过尽可能详尽列举待改进、发明、设计的产品的各种特性，以全面系统解剖产品，把问题化整为零寻求突破的一种创新方法。

2.形态分析法

按材料分解、工艺分解、功能分解、形态分解、成本组成分解，把待解决问题分解成各个独立的要素，然后用图解法将要素进行排列组合，从许多方案的组合中找出最优解，便是形态分析法，也称“形态学矩阵”。

3.类比创新技法

通过比较事物之间在局部上的相同或相似而把性质上不同的事物联系起来的一种思维方法，以便类比推理，有多种形式，如形象类比、关系类比、结构类比、功能类比等。

4.头脑风暴法

一般是通过一种特殊的小型会议，使与会人员围绕某一课题相互启发、激励，取长补短，引起创造性设想的连锁反应，产生众多的创造性成果。

5.对智力激励法的改进

人们在运用智力激励法进行创造的过程中,根据自己的情况对该技术进行了改进。例如635法、MBS法、CBS法以及KJ法。

6.综摄法

在一个主持人召集下,由数人构成一个小组,小组成员的专业范围较广、知识背景不同。会议主题往往是把具体问题提炼后的抽象、简单词汇,运用引喻、类比以及玩弄词义等可操作性的方法,调动小组成员的潜意识,把不熟悉的东西变为熟悉的东西,或把熟悉的东西变为不熟悉的东西,并在相互启发下提出创新的思路。

7.六项思考帽法

运用水平思考法,仅允许思考者在同一时间内只做一件事情,强迫思考者将逻辑与情感、创造与信息等区分考虑,戴不同帽子使人们依次对问题的不同侧面给予足够成分的考虑。最终得出事物的全方位“彩色”答案,每一顶帽子的颜色与其功能是相关的。

#### 4.2.1.3 对产品界面的认知

对产品界面的认识的例子如图4-3和图4-4所示。

图4-3 照相机界面认知

图4-4 车门界面认知

#### 4.2.1.4 作业

通过讨论,利用创新设计方法,对现有产品的设计提出改进方案。

### 4.2.2 项目二 产品设计程式认知

#### 4.2.2.1 目的及要求

1)通过实验,学生能够对“产品设计(一)”课程所涉及的有关设计程式有进一步的理解和掌握,根据认知产品总结产品设计程序。

2)根据课程内容及参观的产品探究产品设计程序。

3)通过实验,学生能够熟悉机电产品造型设计步骤。

4)启发学生创新意识。

#### 4.2.2.2 产品设计程序

产品设计一般是按照如图4-5所示的程序来进行的。

下面以摩托车设计为例,来介绍产品设计的程序。

1.接受项目,制订计划

一般包括项目可行性报告和项目总时间表两个方面。

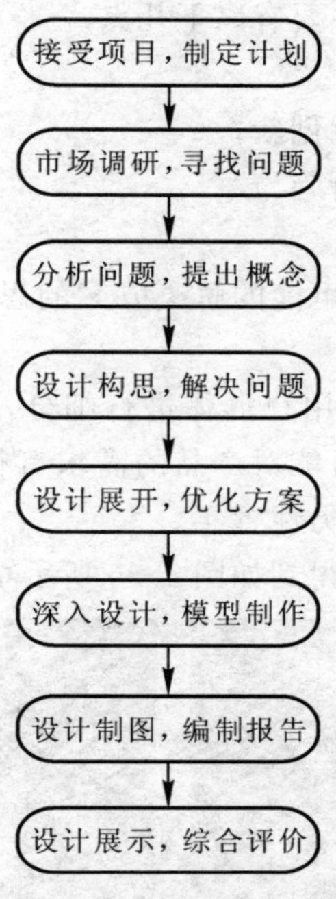


图 4-5　产品设计程序

项目可行性报告是使设计方对业主有深入的了解，以明确自己实施设计过程中可能出现的问题和状况。项目总时间表是根据时间要求，制定一个时间进程计划。

摩托车设计进程安排如下：

1)5 000 字符以上的外文翻译和 2 000 字左右的开题报告　2008.3.3—2008.3.17

2)两轮摩托车的造型调研分析报告　2008.3.18—2008.3.26

3)两轮摩托车造型创新设计方案(手绘)　2008.3.27—2008.4.6

4)确定方案的油泥模型制作　2008.4.7—2008.4.27

5)对油泥模型进行数据采集　2008.4.28—2008.5.4

6)两轮摩托车的效果图制作　2008.5.5—2008.5.18

7)两轮摩托车的标志、铭牌设计　2008.5.19—2008.5.25

8)综合评价及展板设计　2008.5.25—2008.6.06

2.市场调研，寻找问题

通过对品种的调研，搞清楚同类产品市场销售情况、流行情况，以及市场对新品种的要求；现有产品中的内在质量、外在质量所存在的问题；消费者不同年龄组的购买力，不同年龄组对造型的喜好程度，不同地区消费者对造型的好恶程度；竞争对手产品策略与设计方向，包括品种、质量、价格、技术服务等；对国内外期刊、资料所反映的同类产品的生产销售、造型以及产品的发展趋势的情况也要尽可能地收集。

在摩托车的设计中通过对国内外摩托车造型、摩托车历史等方面的调研，对摩托车现有市场进行一定的了解。

3.分析问题，提出概念

经过调研，对所收集的资料进行分析，发现问题所在。了解摩托车的设计趋势，明确设计概念。摩托车造型设计的发展趋势主要有以下几点：

1)高科技与时尚共存；

2)外形在注重人情味的同时注重创新；

3)节能环保的绿色设计是大势所趋。

4.设计构思,解决问题

构思,是对既有问题所作的许多可能的解决方案的思考。根据我们前面所说的创新设计方法来进行草案的设计。

在进行草案设计之前,可以先对用户群体进行细分,通过前面调研可以看出不同年龄阶段、不同性别、不同生活环境的用户可能对产品的需求有很大的差别。因此,摩托车的设计将市场定位为年龄在 18～35 岁的乡镇男性青年。

再对他们的需求进行分析,可以得到如图 4-6 所示草案。

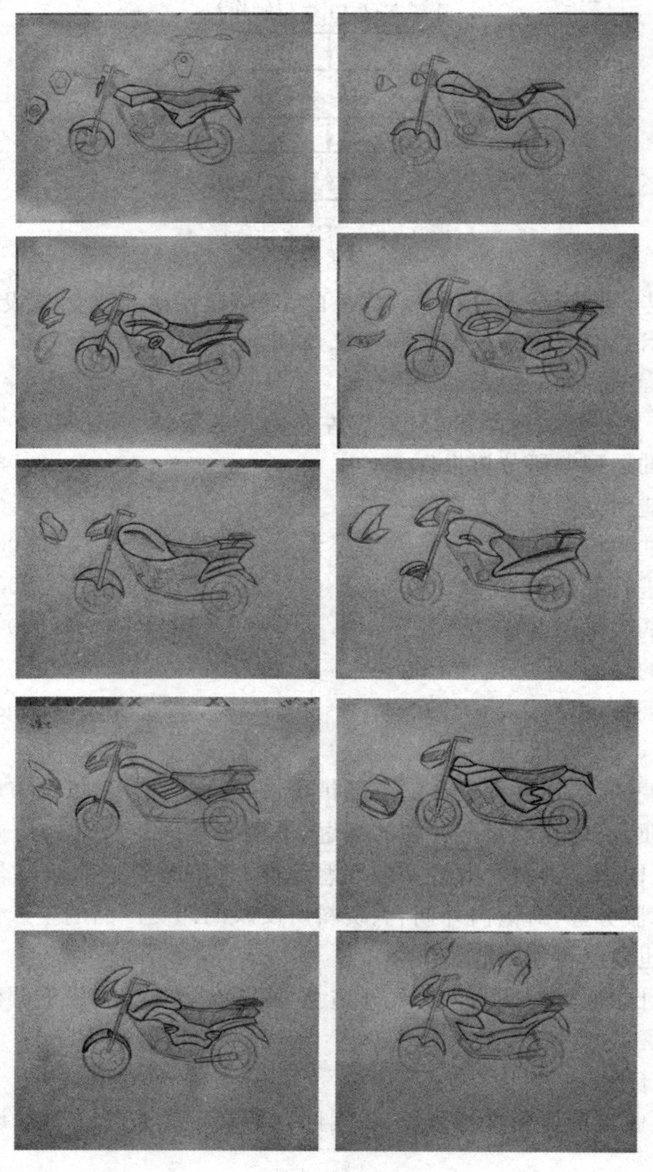

图 4-6　设计草图

5. 设计展开，优化方案

在商品化设计中，产品方案的决策应联系协调各相关部门进行分析评价。首先决定出最符合市场前景的方案进行生产、试销，然后再投放市场。由于条件的限制，请了一些人员而非公司各部门人员进行了评价。在这里首先采用适合于多方案的 α·β 法，评价后淘汰 7 个方案，留下 3 个备选方案。然后再分别采用名次计分法和总分计算法来在剩余的 3 个方案中进行评价。

最终得到方案三为优化方案。

6. 深入设计，模型制作

在模型制作中，我们选择用油泥做模型。

油泥模型是指在产品造型设计的过程中，依照胶带图所贴出的产品的造型设计方案的轮廓线，用工业油泥加热后覆盖在泡沫等轻质材料切割成的内芯上，使用相应的油泥刮刀、刮片等工具制作出来的产品造型设计方案的实体模型。

做出模型如图 4 - 7 所示。

图 4 - 7　制作的油泥模型

7. 设计制图，编制报告

设计报告的制作要全面、精练，不可拖泥带水。一般包括以下内容：封面、目录、设计计划进度表、设计调查、分析研究、设计构思、设计展开、方案确定、综合评价。

8. 设计展示，综合评价

设计报告，在有些情况下要做成设计展示版面。因此在设计完成后，一般都做一些铭牌设计、展板设计等。对设计的综合评价方式一般从两个原则入手：①该设计对使用者、特定的使用人群及社会有何意义？②该设计对企业在市场上的销售有何意义？

摩托车设计的展板如图 4 - 8 所示。

图 4 - 8　展板制作

#### 4.2.2.3 作 业

产品程式化设计的步骤是什么？确定一产品，完成产品造型的设计步骤。

## 4.3 工业设计(二) Industry design（Ⅱ）

### 4.3.1 项目一 针式打印机功能分析拆装

#### 4.3.1.1 目的

1)学会根据针式打印机实物，对实物进行功能分解，分析针式打印机功能树，勾画功能结构图。

2)根据功能结构图最终得出的解，确定针式打印机的创新设计原理方案。

3)根据实物及对最优解的判定，决策出最佳原理方案。

#### 4.3.1.2 设备和工具

起子、钳子、扳手，铅笔、橡皮、草稿纸(学生自备)。

#### 4.3.1.3 功能分析原理

功能分析是从对产品结构的思考转为对它功能的思考，从而做到不受现有结构的束缚，以便形成新的设计构思，提出创造性方案。

在功能分析中，要首先将设计任务抽象化，确定总功能，抓住本质；再将总功能分解为分功能直至功能元；扩展思路，寻求各功能元的解；将所有的原理解组合，形成多种原理设计方案；决策出最佳的原理设计方案。

#### 4.3.1.4 针式打印机组成与功能说明

针式打印机的基本工作原理是利用机械和电路驱动原理，使打印针撞击色带和打印介质，进而打印出点阵，再由点阵组成字符或图形来完成打印任务的。打印机在联机状态下，通过接口接收 PC 机发送的打印控制命令、字符打印或图形打印命令，再通过打印机的 CPU 处理后，从字库中寻找与该字符或图形相对应的图像编码首列地址(正向打印时)或末列地址(反向打印时)，如此一列一列地找出编码并送往打印头驱动电路，激励打印头使针式打印机打印。

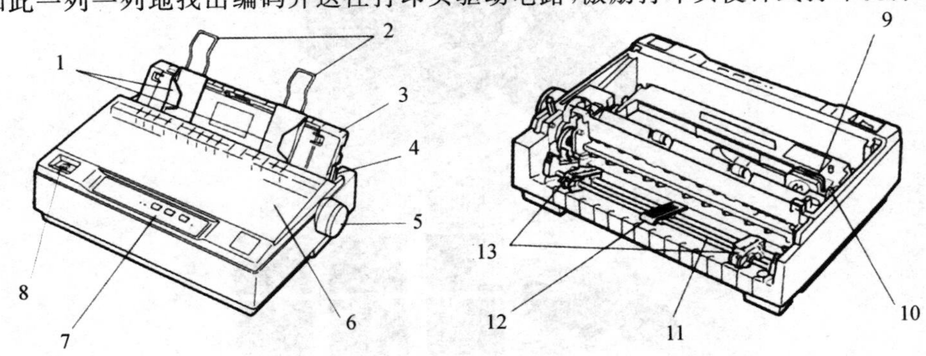


图 4-9 针式打印机结构

1—导轨； 2—钢丝托纸架； 3—导纸器； 4—过纸控制杆； 5—卷轴旋钮； 6—打印机盖； 7—控制面板
8—电源开关； 9—打印头； 10—纸厚调节杆； 11—托纸器； 12—中央托纸块； 13—链齿

1. 针式打印机的结构(见图 4－9)

2. 针式打印机的工作原理说明

针式打印机的结构图如图 4－10 所示。

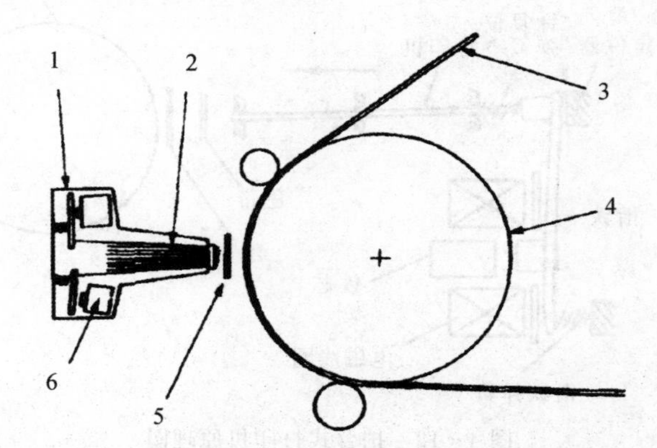


图 4－10　针式打印机的结构图

1—打印驱动部分；　2—打印针；　3—纸；　4—字辊；　5—色带；　6—磁力线圈

针式打印机的成字过程是由打印头里的打印针适时击打色带形成点阵，再由点阵构成文字或图像。打印头中的每一根打印针在打印机电脑的控制下有序地击打色带，在打印纸上留下点阵，这些点阵构成文字或图像。

3. 针式打印机的功能说明

下面从针式打印机的主要功能来分解整个针式打印机结构。

打印机的机构由 4 个部分构成：打印头、字车、色带机构、输纸机构，如图 4－11 所示。

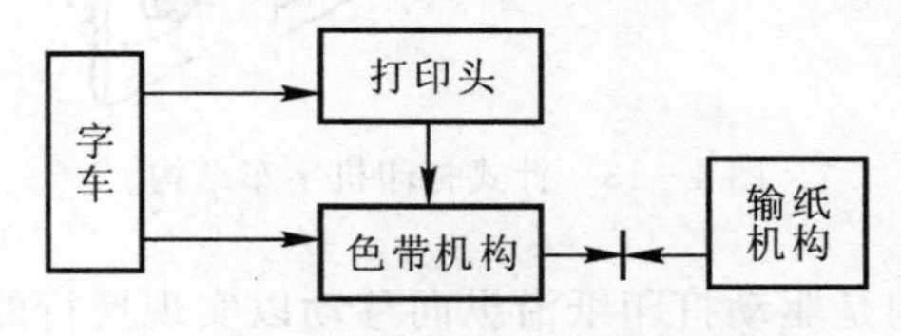


图 4－11　针式打印机结构示意图

1)打印头。打印头即印字机构，它是成字部件，由若干根打印针和相应数量的电磁铁组成，其中电磁铁可驱动打印针完成击打动作；打印头是打印机打字的最终执行机构。打印机经过一系列的数据处理，以及一系列的机械运动之后，由打印头上的打印针完成打印任务。

打印头的打印针的出针方式分为拍合式、储能式、螺管式、音圈式和压电式等。下面详述拍合式打印头的打印原理，如图 4－12 所示。

在图 4－12 中，打印机靠弹簧顶起。当线圈通电时，衔铁受磁力吸引向下冲击与铁芯吸合，带动打印机冲出打印头，把色带和打印机一同撞在打印辊上，使纸上留下一个色点。当线圈上的电压消失时，衔铁在弹簧的作用下回到原位，打印针缩回打印头内。

2)字车。该机构利用步进电机及齿轮减速装置，由同步齿形带来带动字车横向运动。针式打印机字车结构如图 4－13 所示。字车运载打印头沿导轨左右移动，完成打印任务。其运行的动力来自步进电机，其传动装置有挠性齿型皮带传动，少数是刚性齿条传动。

通用打印机字车的取向都是水平安装的卧式，有几款经济型的是斜立式，其他为了便于打印件平推平送，都取立式，打印头是垂直向下的。

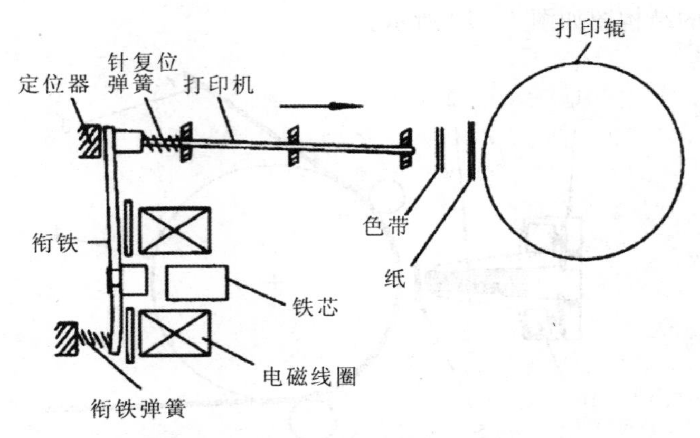


图 4-12　拍合式打印机原理图

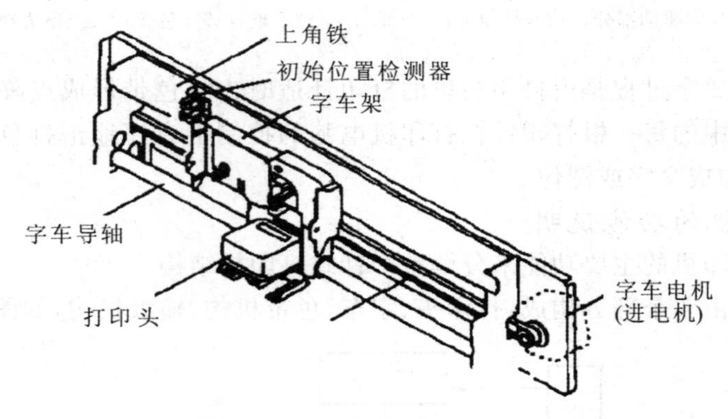


图 4-13　针式打印机字车结构

3)输纸机构。输纸机构是驱动打印纸沿纵向移动以实现换行的机构。

针式打印机的输纸机构一般分为摩擦输纸和齿轮输纸方式，前者适用于无输纸孔的打印纸；后者适用于有输纸孔的打印纸。在打印头完成一行打印后(不管字符多少)，走纸机构将马上完成一行或多行走纸。

输纸机构的动力来源于步进电机，通常将进纸方式分为单页进纸和连续进纸。输纸机构一般由链轮、输纸调节杆、打印辊、打印辊齿轮、惰轮(偏心轮)、压滚以及输纸轴齿轮组成，如图4-14所示。

4)色带机构。在针式打印机中普遍采用单向循环色带机构，打印头左右运动时，色带驱动机构驱动色带向左运动，既可改变色带受击部位，保证色带均匀磨损，延长色带使用寿命，又能保证打印字符颜色深浅一致。色带常用涂有黑色或蓝色油墨的带状尼龙或薄膜制成。

色带的动力除了少数打印机是由专用色带电机带动外，其他的打印机动力多是字车的转矩驱动。由于色带总是向一个方向运动，而字车则是左右移动的，所以色带机构中有换向齿轮。如图4-15所示，当字车由左向右移动时，各齿轮转动顺序如下：色带惰轮→换向齿轮→行星齿轮→色带齿轮轴(如黑色箭头所示)。当字车由右向左移动时，各齿轮转动顺序如下：色带惰轮→换向齿轮→行星齿轮①→行星齿轮②→行星齿轮③→行星齿轮④→色带齿轮轴(如中空箭头所示)。

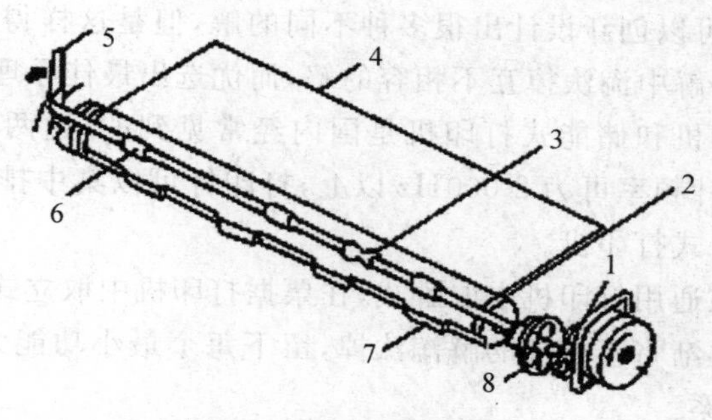


图 4－14　输纸机构结构

1—输纸轴齿轮；　2—链轮；　3—纸；　4—输纸调节杆；　5—压滚；　6—打印辊；　7—打印辊齿轮；　8—偏心轮

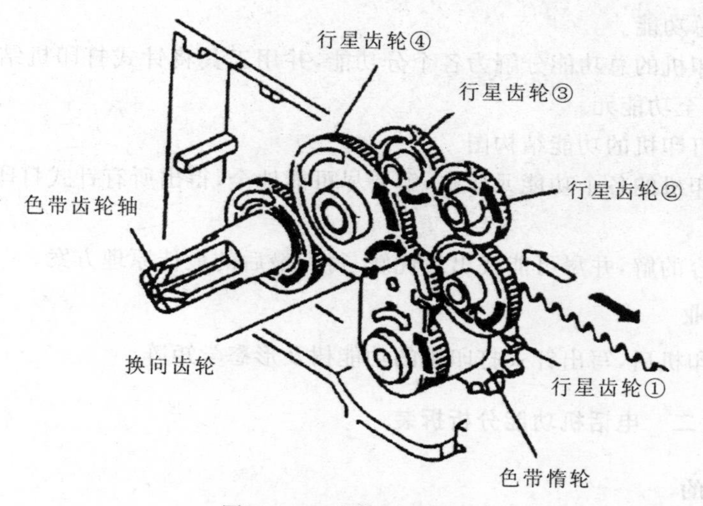


图 4－15　色带机构结构

4. 针式打印机的创新设计原理

根据上面的拆分，我们可以把针式打印机从功能角度分解为子功能直至功能元，然后对功能元求解，以期得出针式打印机的创新原理方案。这里就要利用形态学矩阵的方法来进行创新设计(见表 4－2)。

表 4－2　针式打印机的创新设计原理方案

| 功能元 | 功能元解 | | | | |
|---|---|---|---|---|---|
| | 1 | 2 | 3 | 4 | 5 |
| 打印头 | 拍合式 | 储能式 | 螺管式 | 音圈式 | 压电式 |
| 字车 | 卧式 | 斜立式 | 立式 | | |
| 色带盒 | 单向循环 | | | | |
| 输纸机构 | 摩擦输纸 | 齿轮输纸 | | | |

5.最佳原理方案及其决策

这样一共可以得到 5×3×1×2=30 种解。

虽然用形态学矩阵可以创新设计出很多种不同的解,但是这样得到的很多解也是不合理的解,所以还需要在这些解中淘汰掉互不相容的解,而优选出最佳原理方案。

打印头:拍合式打印机和储能式打印机是国内经常见到的,这两种打印头的结构大致相同。拍合式打印机的出针频率可为 2 000Hz 以上,打印针可以集中排列,无需弯曲,打印头的体积小,故优先选用拍合式打印机。

字车:一般打印头在通用打印机中取卧式,在票据打印机中取立式,因此优先选卧式字车。

同上面两例一样,逐渐将不合理的解淘汰掉,留下每个最小功能元的最优解,再由这些最优解组合成最佳原理方案。

#### 4.3.1.5 方法与步骤

1)仔细观察针式打印机的结构,并分析其原理,根据针式打印机的工作原理,用黑箱法确定针式打印机的总功能。

2)将针式打印机的总功能分解为各个分功能,并用工具将针式打印机结构根据各个分功能分解开,并分解至功能元。

3)勾画针式打印机的功能结构图。

4)对针式打印机的各个功能元作解,解要尽可能地全,得出所有针式打印机的创新设计原理方案。

5)淘汰不相容的解,并尽可能找出最优解,得到最后的最佳原理方案。

#### 4.3.1.6 作业

拆装针式打印机后,写出针式打印机的功能树及形态学矩阵。

### 4.3.2 项目二 电话机功能分析拆装

#### 4.3.2.1 目的

1)学会根据电话机实物,对实物进行功能分解,分析电话机功能树,勾画功能结构图。

2)根据功能结构图最终得出的解,确定电话机的创新设计原理方案。

3)根据实物及对最优解的判定,决策出最佳原理方案。

#### 4.3.2.2 设备和工具

同项目一。

#### 4.3.2.3 功能分析原理

同项目一。

#### 4.3.2.4 电话机组成与功能说明

1.电话机的基本工作原理

电话机的工作原理是利用电磁感应进行信号转换。话筒内有个振动膜,说话时声音是机械波,会使振动膜震动,产生感应电流,根据声音的大小不同,振膜震动情况不同,因此通过的电流大小也不同。电流通过处理,在另一台电话中用仪器把点信号转化回声音信号。而这仪器可以理解为绕有线圈的永久磁铁,电流通过线圈时产生感应磁场,吸引磁铁中的薄铁片产生振动，发出声音。

2. 电话机的功能说明

下面从电话机的主要功能来分解整个电话机结构。

电话机由 4 个部分构成：话筒、听筒、数据传输接头和电话号码输入。部件拆分如图 4－16 所示。

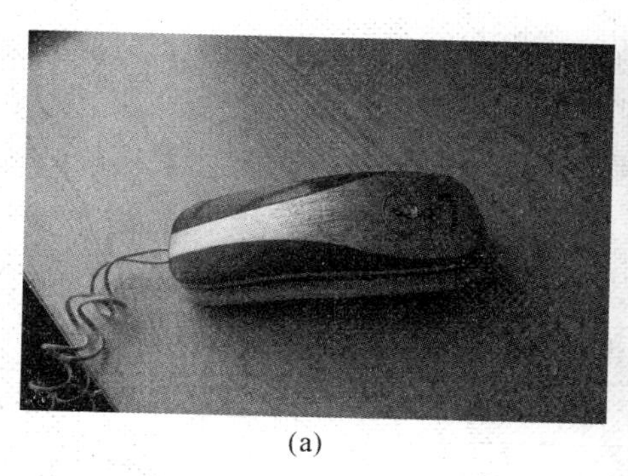

(a)

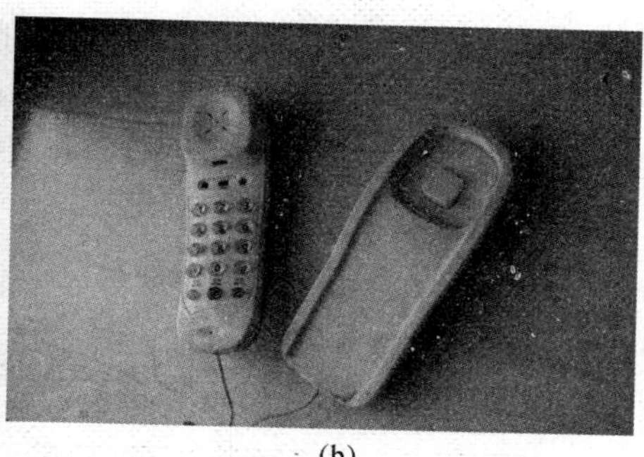

(b)

图 4－16　电话机结构

#### 4.3.2.5　方法与步骤

1)仔细观察电话机的结构，并分析其原理，根据电话机的工作原理，用黑箱法确定电话机的总功能。

2)将电话机的总功能分解为各个分功能，并用工具将电话机结构根据各个分功能分解开，并分解至功能元。

3)勾画电话机的功能结构图。

4)对电话机的各个功能元作解，解要尽可能地全，得出所有电话机功能元的创新设计原理方案。

5)淘汰不相容的解，并尽可能找出最优解，得到最后的最佳原理方案。

#### 4.3.2.6　作业

根据功能分析法分析电话机的功能，写出电话机的功能树及形态学矩阵，并设计一款电话机的造型。

## 4.4　工业设计(三)　Industry design (Ⅲ)

### 4.4.1　项目一　瓦楞纸箱的设计与制作

#### 4.4.1.1　目的

1)掌握瓦楞纸的性能和特点；

2)能够进行瓦楞纸箱的包装设计；

3)掌握瓦楞纸箱的制作方法与成型。

#### 4.4.1.2　原理

本次实验依照瓦楞纸箱的生产工艺原理进行。普通纸箱的生产工艺如下：

纸板→模切→印刷→模切→压线→切角→接合→瓦楞纸箱。

简单而言，模切和压线是按照箱坯结构图进行切断边、开槽、横压线、纵压线等操作；切角是将箱盖之间的连接部分切开，并切出箱体接舌的操作；接合是采用钉接、黏结和胶带黏结等不同方式，将箱坯接合成纸箱。

#### 4.4.1.3 瓦楞纸的性能和特点

将瓦楞原纸加工成瓦楞形状之后按照一定的方式与箱纸板黏合在一起而形成的多层纸板叫作瓦楞纸板。它由面纸、瓦楞芯、里纸三层构成，具有良好的弹性、强度和挺度。用它制造的纸箱和纸盒包装商品，与传统的木箱、金属桶比较，表现出许多优越性，因此被越来越广泛地应用。

#### 4.4.1.4 瓦楞纸箱的设计

瓦楞纸箱的各种箱型、结构实例与特征如表 4-3 所示。

**表 4-3 瓦楞纸箱的各种箱型、结构实例与特征**

| 特 征 | 箱型代号与结构实例 |
| --- | --- |
| 摇盖纸箱，由一片瓦楞纸板加工而成，经钉、黏或胶带接合成箱后可折叠 | 0201 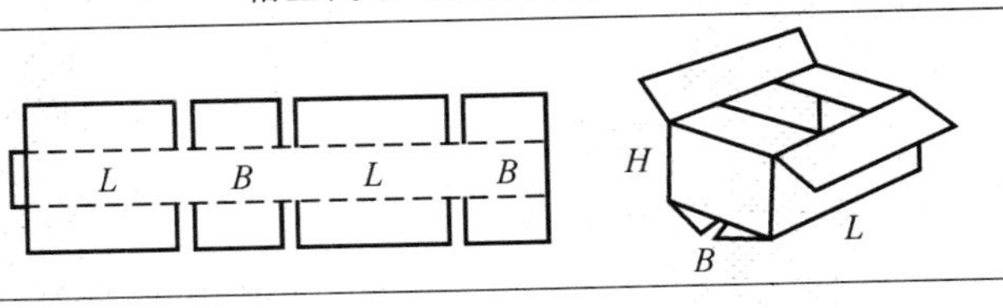 L B L B ; H B L |
| 由两片或三片瓦楞纸板加工而成，一般制成上、下两部分或上盖、箱体、箱底三部分，每部分钉接或黏结成型后套插成箱 | 0312 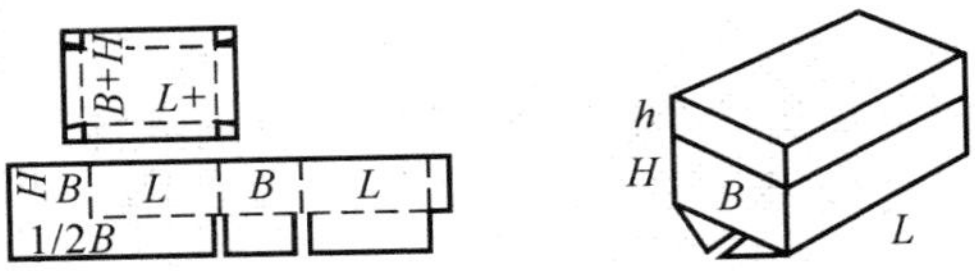 B+H L+ ; H B L B L 1/2B ; h H B L |
| 通常由一片瓦楞纸板制成，无需钉接、黏结成箱。使用时先包装商品，折叠后用包带捆扎即成包装件，适合于在自动生产线上包装成件、成块的商品 | 0443 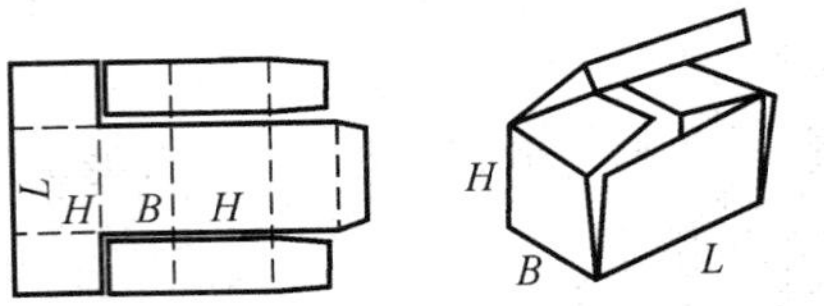 L H B H ; H B L |
| 由一片或两片瓦楞纸板加工成箱体或套盖，然后组建成包装箱 | 0511 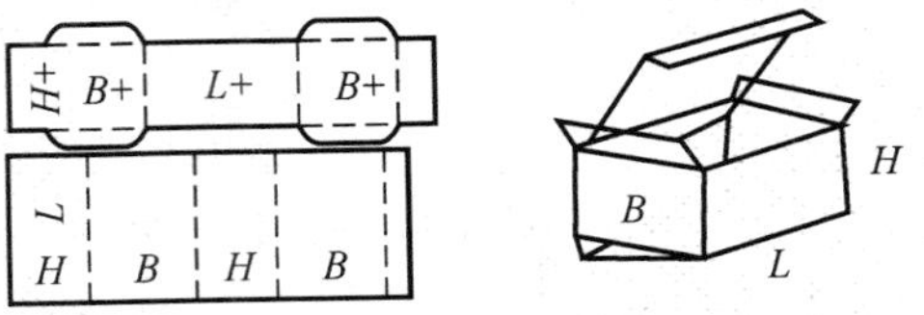 H+ B+ L+ B+ ; L H B H B ; H B L |
| 一般由三片瓦楞纸板制成，一片制成箱体，另两片作为箱的两端板使用；成箱时大都采用普通钉或纸箱钉钉接 | 0601 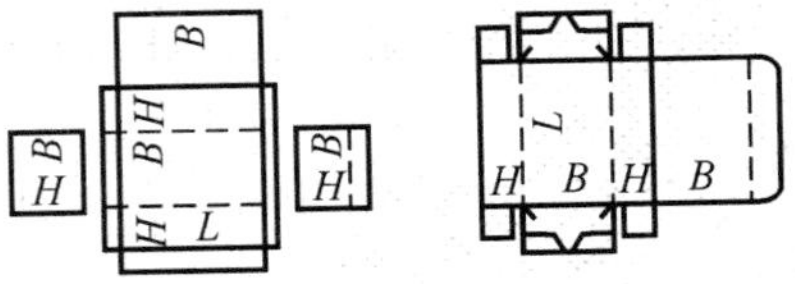 B H ; B B H H L ; B H ; L H B H B |
| 由一片瓦楞纸板加工而成，然后大多用胶黏结成箱。通常在箱体四周开孔，可用于超市销售的蔬菜、水果等的包装 | 748 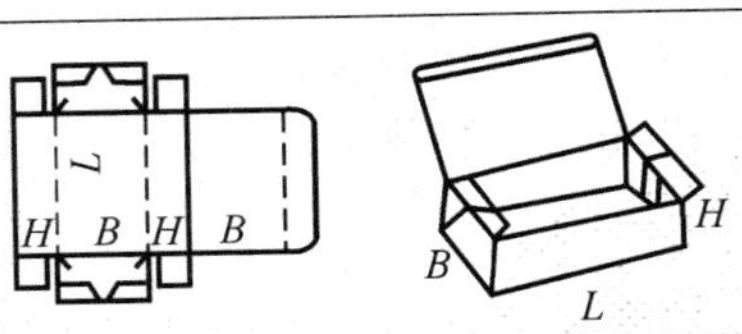 L H B H B ; H B L |

瓦楞纸箱的设计需要考虑的因素有以下几个方面：

1.纸盒的尺寸

瓦楞纸箱的尺寸有 3 种形式：外尺寸、制造尺寸和内尺寸。这 3 种形式都是由纸箱的长度、宽度和高度共同来组成的。因此，具体的表述上分为外长宽高、内长宽高和制造长宽高。外尺寸即为从瓦楞纸箱外部量取的体积的大小，一般作为计算运费的依据和纸箱外部的标示。内尺寸即为瓦楞纸箱支撑成型后从内部量取的面到面的距离，内尺寸的确定是根据内装物的体积大小确定的，一般情况下，内尺寸略大于内装物的体积。制造尺寸即根据纸箱内尺寸换算出的生产工艺尺寸。

外尺寸、内尺寸、制造尺寸三者之间的关系为

外尺寸＝内尺寸＋瓦楞纸板厚度×瓦楞纸板的层数

制造尺寸＝内尺寸＋瓦楞纸板厚度÷2×瓦楞纸板的层数

在这两个关系式里，“瓦楞纸板的层数”指的是纸箱折叠起来以后，箱身和摇盖重合的地方的层数，这个层数因长、宽、高而异。

设计瓦楞纸箱的大小首先要进行内尺寸的设计，因为内尺寸比较容易确定，它是以内装商品的实物测量或者实物组合排列计算的外尺寸来确定的。内装物的排放应该尽量节约空间，并合理考虑支撑力的因素。

2.接盒方式

1)钉接成箱。用钉箱机将纸箱搭接处用铁钉铆合起来，是常用的一种纸箱接合方式。它操作简单、设备成本低，适合小厂使用。

2)黏结成箱。黏结成箱常用聚醋酸乙烯乳胶作为黏合剂，采用自动折叠式粘贴机，由上胶、折叠、贴压风干等部分组成，生产效率高，劳动强度低。

3)胶带接合成箱。胶带接合成箱时箱坯不需要设置搭接舌，将箱体对接后，用强度较高的增强胶带粘贴即可。箱内、外表面平整，密封性好。批量不大时可手工操作。

#### 4.4.1.5　瓦楞纸盒的制作步骤

1)了解内装物尺寸；

2)选择瓦楞纸箱箱型，查阅并选择适当瓦楞纸箱；

3)选择适当瓦楞楞型；

4)计算瓦楞纸箱内尺寸；

5)计算瓦楞纸箱制造尺寸；

6)绘制瓦楞纸箱箱坯图纸；

7)准备瓦楞纸板；

8)使用工具裁切与压线；

9)接合成型。

#### 4.4.1.6 实验报告范例

## 实验报告范例

| 实验名称 | | 瓦楞纸箱的设计与制作 | | | |
|---|---|---|---|---|---|
| 班 级 | | 姓 名 | | 学 号 | |
| 实验日期 | | 成 绩 | | 指导教师 | |

### 一、瓦楞纸箱的设计方案

1)内装物尺寸(长 $L$、宽 $B$、高 $H$)。

$L$=250 mm； $B$=180 mm； $H$=110 mm

2)画出箱型图(见图 4-17)。

3)计算 3 种形式的尺寸(长、宽、高)(mm)。

外尺寸： $L=410+2\times5=420$ mm

$B=340+2\times5=350$ mm

$H=350+4\times5=370$ mm

制造尺寸： $L=410+0.5\div2\times20=415$ mm

$B=340+0.5\div2\times20=345$ mm

$H=350+0.5\div2\times40=360$ mm

内尺寸： $L=250+2\times80=410$ mm

$B=180+2\times80=340$ mm

$H=110+2\times12=350$ mm

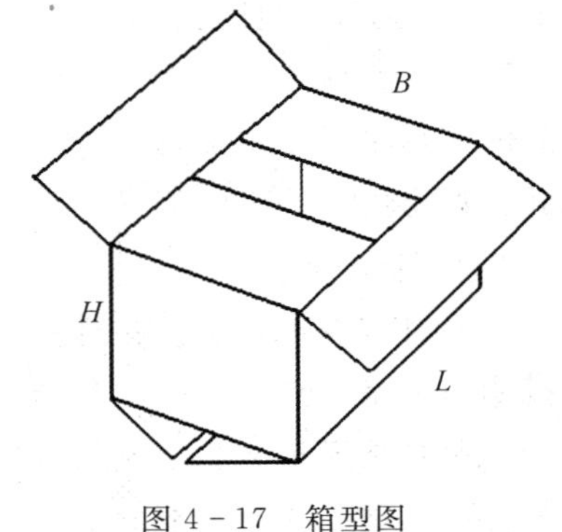

图 4-17 箱型图

4)绘制出箱坯结构图(按比例缩小)(见图 4-18)。

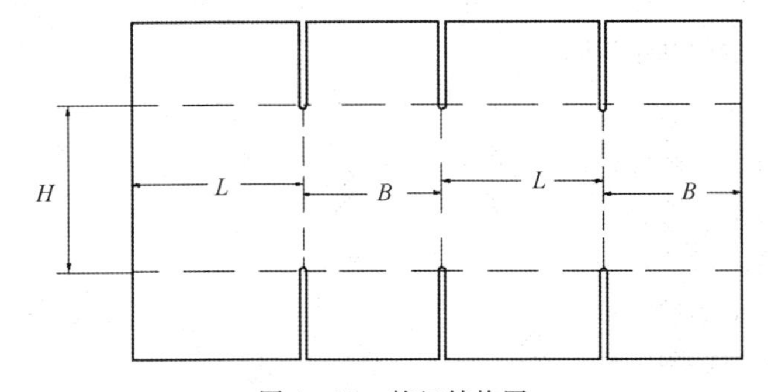

图 4-18 箱坯结构图

### 二、设计与实验心得

通过这次实验，我了解了外包装制作的流程与规律。外包装的尺寸一定要依据内装物的尺寸来决定，不能太大也不能太小。大了会在泡沫填充时浪费泡沫材料；小了则会在泡沫填充时泡沫溢出，造成包装效果差，甚至失败。

而且，当设计纸箱时，也不像想象中那么简单，而是有严格的章法可循。如果尺寸设计不合理，那么在纸箱成型时可能会合不拢或边对不上。

当填充泡沫时，要先填充四边，最后中间，而且在填的时候要均匀、快速。

由于是小组合作，合理的分工是非常重要的，相互之间更要协作才能做出合格的作品。

### 4.4.2　项目二　泡沫塑料包装材料的成型

#### 4.4.2.1　目的

1)使学生能掌握泡沫塑料的成型原理；

2)使学生能了解现场发泡系统的结构、功能以及工作原理；

3)通过动手操作，学生能够了解现场发泡的成型过程。

#### 4.4.2.2　设备和工具

1)现场发泡系统(发泡机)及原料；

2)待包装的产品；

3)产品的外包装(纸箱)。

#### 4.4.2.3　泡沫塑料包装材料的成型原理

1.泡沫塑料

泡沫塑料是以树脂为基础制成的内部具有无数微小气孔的塑料制品。泡沫塑料的通性是质量轻，比同种塑料要轻几倍甚至几十倍，由于有无数小孔，因此不易传热、能吸音、绝缘、防震等。泡沫塑料可作包装材料、绝热材料、吸音材料、过滤材料、室内装饰材料、浮漂物和绝缘材料等。

泡沫塑料的分类方法有很多种。按照气孔结构不同，可分为开孔泡沫塑料和闭孔泡沫塑料；根据泡沫塑料的力学强度，可分为硬质、半硬质和软质三种；按照泡沫塑料的密度，又可以分为低发泡、中发泡和高发泡。

应用现代技术，几乎可将各种热固性和热塑性塑料制成泡沫塑料，但是通常用于制造泡沫塑料的树脂主要有PVC，PS，PE，PP，聚氨基甲酸酯，聚甲醛，环氧树脂等。

2.发泡剂

泡沫塑料的发泡方法有屋里发泡、化学发泡和机械发泡3种，它们的共同点是把制造泡沫塑料的组成物，形成具有一定黏度范围的液体或塑性状态物质，然后借发泡剂或机械搅拌的作用，制成多孔的泡沫塑料，因此发泡剂就是指能够产生泡孔结构的固体、液体和气体，或是几种物料的混合物。

3.发泡方法

物理发泡法具有操作毒性小、发泡原料成本低、发泡剂无残余等特点。常采用的为惰性气体发泡法和低沸点液体发泡法。惰性气体发泡法即将惰性气体在较高的压力下压入熔融聚合物或糊状复合物中，然后降低压力，升高温度使发泡；而低沸点发泡法是把低沸点液体压入聚合物中，或在压力下把液体溶于聚合物颗粒中，然后加热到液体的沸点，借气化所产生的气体使聚合物发泡。

化学发泡法也可分为两种，一种是利用化学发泡剂受热分解产生气体使聚合物发泡，另一种是利用多组分原料相互反应放出惰性气体使聚合物发泡。本实验采用的是化学发泡法，两种原料分别是乙氢酸酯（黑桶中的原料，通常称为 A 料）和聚醚（蓝桶中的原料，通常称为 B 料），发泡剂是水。采用这样的原料和发泡剂，生产的过程安全无毒害，生产出的发泡体防震性能好，是许多出口产品必须使用的泡沫塑料包装。

机械搅拌发泡是利用强烈的机械搅拌，将空气卷入树脂的乳液、悬浮液或溶液，使其形成均匀的泡沫物，而后再通过物理或化学方法使其稳定而成为泡沫塑料。

4. 成型类型

泡沫塑料包装的成型类型可分为两种，一种是模具成型包装方法，一种是现场发泡包装方法。

模具成型包装方法需要预先制定模具，然后注入发泡液；而现场发泡方法不需要模具，只要直接利用薄膜和纸箱就可以完成包装过程。成型效果的区别是显而易见的，通过模具成型包装方法完成的发泡垫形状规则、整齐，表面平展，适合于大批量生产、成本较高的产品。现场发泡方法不需要制作模具，以节约成本。生产出来的包装垫具有随意性，适合于形状比较复杂、非批量生产的物品。

无论是模具成型包装方法，还是现场发泡方法，生产出来的包装垫都可以起到固定产品、缓冲压力，有效地保护产品不受损坏的作用。

#### 4.4.2.4 现场发泡系统的结构

现场发泡系统中所采用的发泡机如图 4－19 所示。

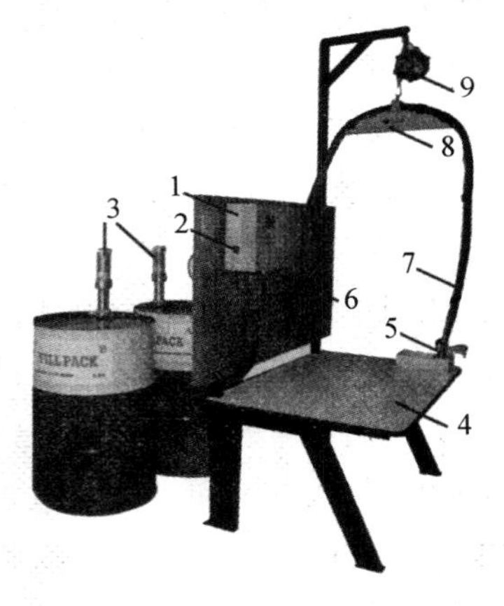


图 4－19 发泡机

1—多功能控制器； 2—电源开关； 3—液泵； 4—标准型工作台； 5—发泡混合枪； 6—气源开关
7—三层金属输液软管； 8—输液管悬挂架； 9—平衡器

#### 4.4.2.5 现场发泡的方法与步骤

1. 系统的使用准备

1）打开空气泵，观察操作台后面的气压总表，当压力达到 6kg 左右时，可以正常操作。打开气源开关，然后再打开电源开关。

2)设置界面。主要是设置发泡液需要达到的温度,通常为65℃或者70℃。

SET:参数设置

RUN:运行

STOP:停机

3)观察发泡过程与发泡体。待温度显示达到所设置的温度时,打出少许发泡液,进行3个方面的观察。

首先观察喷射出的发泡液的颜色。正常的颜色应该是乳白色,如果颜色发黑,说明A料枪嘴堵塞;如果颜色过于发白,说明B料枪嘴堵塞;此时要用通针疏通枪嘴,并用枪嘴清洗剂进行洗涤。

其次观察发泡过程的速度,如果膨胀速度太慢,有可能是温度较低或者原料比例不当。温度较低的情况下可参考步骤3)对温度进行调整。

最后观察发泡体是否为乳白色,泡孔是否均匀。若发泡体颜色深、质地比较脆、无弹性,用手搓的时候会掉渣,则说明A料过多而B料较少,需要将A料的减压阀向"一"的方向调节,或者将B料的减压阀向"+"的方向调节;若发泡体质地较软,向下塌陷,则说明B料过多而A料较少,需要将B料的减压阀向"一"的方向调节,或者将A料的减压阀向"+"的方向调节。调节的幅度为每次约0.5kg,调节之后再次打出少许发泡液进行观察,如果发泡不理想,继续调节,直至可以形成泡孔均匀的乳白色的发泡体。

2.现场发泡的过程

1)根据包装箱的大小准备薄膜(通常为两张)。

2)将薄膜均匀地铺在箱底,注意将印刷有字体的一面朝上放置。

3)在薄膜上打出适量的发泡液,在发泡液发泡的过程中,迅速地将薄膜四周向中间包起,直到包住所有正在形成的发泡体。然后将需要保护的产品平放在发泡体上,向下压至产品的近一半陷入发泡体内。

4)将另一张薄膜继续铺在箱子里,注意使其尽量紧贴产品;打出适量的发泡液,迅速包起薄膜,并将箱盖压紧,即可完成包装。

3.系统的关闭

1)关闭电源;

2)关闭气源;

3)将枪头浸泡在浸枪盒中。

**4.4.2.6 操作的若干注意事项**

1)开机的时候先开气源后开电源,关机的时候先关电源,后关气源。

2)放置薄膜的时候注意保持印刷有字体的一面向上。

3)较长时间不使用的时候,要及时关机。

4)此系统不应放于密闭、狭窄或无新鲜空气及通风设备的地方。

5)液体的泡沫可能会接触到操作人员的衣服、皮肤或工作台表面,泡沫在逐渐硬化时会黏在表面并促使其轻微地变色。一般情况下,应等泡沫硬化后再予去除。如误入眼睛,迅速以大量清水冲洗眼睛15min并及时就医。可能暂时造成疼痛及不适,但不会造成永久性受伤。

6)严禁将发泡枪对准人体!

#### 4.4.2.7 实验报告范例

## 实验报告范例

| 实验名称 | 泡沫塑料包装材料的成型 | | | | |
|---|---|---|---|---|---|
| 班　级 | | 姓　名 | | 学　号 | |
| 实验日期 | | 成　绩 | | 指导教师 | |

### 一、现场发泡系统的结构及其对应功能

| 结　构 | 功　能 |
|---|---|
| 多功能控制器 | 控制发泡机运行的开始与停止,控制加热温度和发泡时间 |
| 电源开关 | 打开或关闭发泡机总电源 |
| 液泵 | 从原料桶中抽取原料输送到输液软管中 |
| 标准型工作台 | 用于操作的台面 |
| 发泡混合枪 | 两种原料经过加热以后,在混合枪中混合、喷射出去,急速膨胀,固化为一种有弹性的缓冲包装材料 |
| 气源开关 | 打开或关闭气源 |
| 三层金属输液软管 | 内部的加热管可加热原料,并将热原料输送到发泡混合枪 |
| 输液管悬挂架 | 固定、悬挂输液管 |
| 平衡器 | 可以任意角度地拉动输液管和发泡混合枪到一定的距离,以便根据产品和场地的具体情况灵活操作 |

### 二、发泡效果不佳时的判断与调整

| 现　象 | 原　因 | 应采取的措施 |
|---|---|---|
| 发泡液颜色接近白色 | B料枪嘴堵塞 | 用通针疏通枪嘴,用洗涤剂进行清洗 |
| 发泡液颜色发黑 | A料枪嘴堵塞 | 用通针疏通枪嘴,用洗涤剂进行清洗 |
| 发泡体膨胀缓慢 | 温度较低或者原料比例不当 | 将温度设置高一点或改变A,B料的配比 |
| 发泡脆,颜色深 | A料过多而B料较少 | 将A料减压阀向"—"方向调节;或将B料减压阀向"+"方向调节 |
| 发泡软,塌陷 | B料过多而A料较少 | 将B料减压阀向"—"方向调节;或将A料减压阀向"+"方向调节 |

**三、现场发泡的操作过程**

1）根据包装箱的大小准备两张薄膜；

2）将薄膜均匀地铺在箱底，注意将印刷有字体的一面朝上；

3）在薄膜上打出适量的发泡液，在发泡液发泡的过程中，迅速将薄膜四周向中间包起，直到包住所有正在形成的发泡体；

4）将需要保护的产品平放在发泡体上，向下压至产品的近一半陷入发泡体内；

将另一张薄膜继续铺在箱子里，注意使其尽量紧贴产品；打出适量的发泡液，迅速包起薄膜，并将箱盖压紧，即可完成包装。

**四、操作中应注意的问题**

1)开机的时候先开气源后开电源，关机的时候先关电源，后关气源。这样做的目的是避免输液管中没有燃料，引起发热线干烧或者短路！

2)放置薄膜的时候注意保持印刷有字体的一面向上，温热的泡沫会和有字的一面紧密合并，而另外一面并不会有这样的效果。

3)较长时间不使用的时候，要及时关机以减少输液软管的受热时间，延长使用寿命。

4)此系统应放于宽敞、空气流通、有通风设备的地方。

5)注意不要将液体的泡沫接触到衣服、皮肤或其他表面。如误入眼睛，迅速以大量清水冲洗眼睛 15min 并就医。

6)严禁将发泡枪对准人体！

## 4.5　造型表现技法　Modeling performance techniques

### 4.5.1　目的

1)熟练掌握刻绘软件的使用；

2)学会用刻绘软件进行各种图形、图案的设计；

3)学会正确使用和调整电脑刻字机；

4)能自行设计、制作各种展板、广告牌等；

5)通过实践，学生能够具备设计与制作平面广告媒体的能力。

### 4.5.2　要求

1)要求学生掌握软件功能，并能够对图库进行创造性重组和设计；

2)要求学生能够灵活运用软件完成一幅广告牌的设计；

3)要求学生能正确操作电脑刻字机，能够根据具体环境调整刀尖压力；

4)要求学生 2～3 人制作一块展板(大小为 A3)。

### 4.5.3　掌握文泰刻绘软件

1)通过上机实践，掌握基础操作，能对文字进行艺术化加工和排版；

2)学习如何绘制图形,能对图形进行复杂的编辑;

3)学会设计展板花边;

4)学习图库的调出和编辑;

5)掌握复杂操作和节点编辑;

6)对表格作各种操作;

7)组合和分割图形;

8)对图形进行"并""交""异或""取差"运算;

9)对文字和图形进行加粗、变空心等特殊效果的处理;

10)对扫描和下载的图像进行转换;

11)掌握文泰刻绘软件的其他操作;

12)熟悉文泰刻绘软件的各种功能,并能灵活、综合运用。

### 4.5.4 设计展板、广告牌

1)确立展板或广告牌的主题,手绘设计效果图;

2)使用文泰刻绘软件,在计算机上设计出此展板或广告牌;

3)综合运用"工业造型与表现技法"课程中学到的原理和法则,力争使设计的展板具有实用价值和审美价值;

4)对文字、图形、背景等要素进行色彩设计,要求色彩搭配协调,具有美感。

### 4.5.5 电脑刻字机刻绘输出

1)了解刻字机的基本构造、联机方法;

2)学会根据刻绘纸的品种、厚度适当调整刀尖的压力;

3)掌握操作计算机和刻字机,刻绘出预先设计的作品的方法;

4)正确操作计算机和刻字机,避免操作不当损坏设备。

### 4.5.6 制作展板、广告牌

1)掌握粘贴文字、图案等的正确方法;

2)将刻绘好的文字、图案等粘贴在展板上。

3)事先对展板进行分区布局设计;

4)要求展板美观、整洁,色彩协调,布局合理,疏密有致。

## 4.6 造型材料与工艺 Modeling materials and processes

### 4.6.1 目的

1)通过实验,对课程所涉及的有关造型材料的特点有进一步的理解和掌握;

2)培养学生对造型工艺特点的认知能力;

3)了解造型工具的特点和使用方法;

4)熟悉常用造型材料的特点及工艺特征;

5)了解常用造型工具的有关特点和使用方法。

### 4.6.2　要求

1)通过参观常用造型材料样品及由其模型制作的产品展示柜,参观并了解常用造型工具;
2)熟悉常用造型工具的特点及用途;
3) 熟悉造型材料及造型工艺。

### 4.6.3　认知内容

1.黏土模型(见图 4－20)

用于模型制作的陶泥是黏土材料的一种,通常是用水调和质地细腻的生泥,经反复砸揉而得,其黏合性强,使用时柔软而不黏手,干湿度适中为宜。陶泥可塑性大,可以根据设计构思自由反复塑造,在塑造过程中可随时添补、削减,充分体现了陶泥材料在塑制过程中的优点。

图 4－20　黏土模型

由于陶泥模型所使用的黏土属水性材料,干燥后易裂、不便保存,一般多用于产品设计时研究性模型的制作或翻制成石膏模型以便保存。

由于陶泥模型材料受气候、温度、湿度的变化影响会产生收缩和变形,所以对尺寸精度有严格要求的设计,通常要求采用质量稳定、塑性较好的陶泥作为模型的塑造材料。因此,陶泥的品质是模型质量好坏的关键,选择质量好的陶泥材料有利于塑造过程的顺利进行。

2.石膏(见图 4－21)

泥模型虽然采用湿布、喷水的方法来保持泥模型的水分和防止杂质的混入,但毕竟这种方法仅能保持短暂时间的不变形,时间一长,水分逐渐消失,仍然会导致模型的收缩变形和干裂。为了使产品模型可以长久保存下去,人们通常采用将泥模型翻制成石膏模型的方法来保存作品,以便长久地保留所塑造的产品形态,同时也可以通过制作石膏模具的方法进行多次复制原形。

由于采用石膏模具的方法翻制产品模型成本低,不需运用太多的工具,操作占地面积小,操作简单,所以一直被广泛地应用于艺术设计、模型制作的领域。在模型成型技法中,这是一种重要的,也是最常用的成型方式。

石膏模具法是石膏成型技法中翻制模型惯用的一种方法。通常,石膏模具是在已塑造完成的黏土或油泥模型母体上,抹上一层脱模剂,在泥塑母体上浇注上一层具有一定厚度的石膏浆,当石膏浆完全凝固时,再取出泥塑的原形,形成中空的石膏模具。

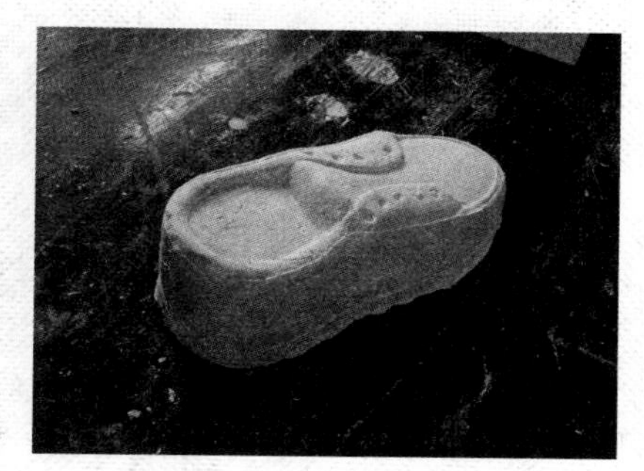

图 4－21　石膏模型

石膏模具与原泥塑模型成为形态上正好相反的阴性模关系。这时泥塑原形已被破坏,而利用阴性的石膏模具来复制保留原来的作品。在阴性的石膏模具型中,浇注石膏浆,待石膏浆凝固成型之后,敲碎阴性的石膏模具,或分片、分部的分开阴性石膏模具,就可以得到复制模型,即石膏的作品原形。

石膏模型能较好地保留设计者原创作品的形态。如果翻制完美,石膏材料能 100%地复原出所设计产品的原形，很好地保留和传递作品的形态。石膏模型有利于保存，假若石膏模具被分片、分部分地剥离，复制出来的石膏模型母体不被损坏，便可以多次复制，复制出多个相同的作品。石膏模型是由石膏粉和适量的水混合后结成的固体物，石膏粉与水的配比值将会影响石膏浆凝固后的强度和密度，如果石膏粉与水的配比比例大，所得的石膏坯块的强度高、密度大，反之则强度低、密度小。

3.泡沫塑料(见图 4－22)

泡沫塑料又称微孔塑料，是由塑料颗粒通过物理加热发泡，或化学方法膨胀发泡形成的塑料制品。泡沫塑料按照密度可以分为硬质和软质两大类，硬质泡沫塑料常用于制作隔热保温材料或隔音防震材料等，软质泡沫塑料常见的是用于制作拖鞋、坐垫、泡沫人造革以及包装材料。工厂生产的泡沫塑料产品的一般成型方法有注射、挤出、模压、浇铸等，这些产品具有质轻、防震、隔热、隔音、耐潮等优点。

模型制作所用的泡沫塑料，最佳选择是硬质泡沫塑料，因为其结构细密均匀，强度较高，加工性好，无论是锯、切、钉，还是打磨，都能保持比较平整光滑的表面，不变形，不收缩。经过表面处理之后，上色均匀有光泽，但是价格相对较贵。对精度要求极高的模型，则宜采用聚甲基丙烯制成的发泡材料，强度高、表面光滑，但价格也非常昂贵。

最常见的聚苯乙烯泡沫塑料是我们最容易得到的泡沫塑料。使用聚苯乙烯泡沫塑料制作模型的优点是材料获取容易，价格低廉，制作模型的成本低。由于它的密度较小，所以很容易切割，能够在较短的时间内完成模型的成型。缺点是表面粗糙，切割和打磨的过程中容易造成材料表面缺失或损伤，无法进行精细的表面处理和上色。这些特点决定了聚苯乙烯泡沫塑料模型不适合制作细节多、造型复杂的产品模型，而常用于形态简单、造型变化不多的产品模型。聚苯乙烯泡沫塑料模型最大的实用意义，是在确定产品设计方案的过程中制作产品草模，迅速获得设计方案的实体形象概念，从而进行进一步的探讨。

图 4－22　泡沫模型

图 4－23　木制模型

4.木模型(见图 4－23)

木材被广泛地用于传统的模型制作中。在传统的机械制造业中大量采用木模型作为铸造用模具。业余模型爱好者使用条状的薄木板做成各种流线型的航天及航海模型，无论把木材用在何处，都能制作出非常精美的作品。

由于木模型对所使用的材料有较高的要求，同时制作木模型需要熟练的技巧和大量的时间，所以在产品模型制作中，通常使用木材来做细致的模型部分，或作为制作产品模型的补充材料，较少用它做结构功能性模型。

在木模型制作中完全使用木材来制作模型可以达到非常精美的效果，但与其他的材料相比，它需要用到各种不同的加工和整饰方法。为了节省时间，增强木模型的表现效果，经常采用木材与其他装饰效果好的表面材料结合使用(如纸材和塑料)。

木质产品的基本加工过程，需要将原料通过木工工具或机械设备加工成构建，再将构建装配成制品，进行表面处理和涂饰，完成作品。基本的操作包括锯割、刨削、凿削等。木材弯曲的基本方法有锯割弯曲、实木软化弯曲、薄木胶合弯曲等。

5. 纸模型(见图 4－24)

纸在生活中到处可见，书籍、笔记本、包装盒、餐巾纸等。对纸的分类有很多种。按照用途，可以分为印刷用纸、包装用纸和特种用纸三类；按照纸张的厚度，可以分为单层纸、双层纸、三层纸等；按照纸张强度和表面的特性，又可以分为卡纸、白板纸、瓦楞纸等。

用纸材制作模型，常见于建筑模型或者室内模型，或者是外观形态比较简洁的产品模型。不是所有的模型都可以用纸来制作，由于纸材无法像石膏或泡沫塑料那样随意雕刻和打磨，因此不适合制作表面凹凸变化太大的产品模型。

图 4－24　线制模型

常见的制作模型的纸材类型有白卡纸、白板纸、箱板纸、瓦楞纸等。在通常情况下，纸模型制作的选料要求有一定的强度和硬度，表面平整光滑，纤维组织均匀，耐折耐弯。

用纸材制作模型，虽然与其他模型制作方法有很大区别，但是由于制作的过程要求准确细致，因此成型后的产品往往精巧雅致，有不同于其他模型的优点。

### 4.6.4　作业

1)了解各种材料在模型制作中的特点；

2)要求会根据造型的不同选择不同的材料。

# 参 考 文 献
# References


[1] Lidwell W, Holden K, Butler J. Universal Principles of Design. Beverly: Rockport Publishers, 2003.

[2] Asensio P. Product Design. New York: Te Neues Publishing Company, 2004.

[3] Norman D. Psychology of Everyday Things. New York: Basic Books, 2002.

[4] Dixon T. The International Design Year Book. New York: Abbeville Press, 2004.

[5] Ulrich K T, Eppinger S D. Product Design and Development. Beijing: Higher Education Press, 2004.

[6] Abraham M. Motivation and Personality. New York: Harper, 1954.

[7] Vecera S P, Vogel E K, Woodman G F. Lower Region: A New Cuefor Figure-Ground Assignment. Journal of Experimental Psychology, 2002, 131(2):194 - 205.

[8] Sofko W W. The Role of Cost-Benefit Analysis in Achieving Results in Special Education. http://www. wildwood. edu/institute/knowledge/results. html.

[9] Jung C G. The Archetypes and the Collective Unconscious. Princeton: Princeton University Press, 1981.

[10] Archer D, Iritani B, Kimes D D, et al. Face-ism: Five Studies of Sex Difference in Facial Prominence. Journal of Personality and Social Psychology, 1983, 45:725 -735.

[11] Newman O. Design Guidelines for Creating Defensible Space. Washington D. C. : Government Printing Office, 1973.

[12] Maslow A H, A Theory of Human Motivation. Psychological Review. 1943, 50:370 - 396.

[13] Erlhoff M. Product Design. Berlin: Feierabend Verlag, Ohg, 2004.

[14] Roqueta H. Product Design. New York: Te Neues Publishing Company, 2003.

[15] "icsid. org" (2011), http://www. icsid. org.

[16] "idsa. org" (2011), http://www. idsa. org.

[17] "idshk. org" (2009), http://www. idshk. org.

[18] "jidpo. org" (2010), http://www. jidpo. or. jp.

[19] "Stiftelsen Svensk Industridesign" (2009), http://www. svid. se.

[20] "GDC. de" (2009), http://www. german-design-council. de.

[21] "design center Stuttgart" (2010), http://www. design-center. de.

[22] "idsa. org" (2011), http://www. idsa. org/idea-2011-gallery.

[23] "iF Design" (2013), http://www. ifdesign. de.

[24] "red dot GmbH & Co. KG" (2003), http://www. en-reddot. org.

[25] "Apple Inc. " (2013), http://www. apple. com.

[26] "frog design inc" (2013), http:// www. frogdesign. cn.

[27] “IDEO inc”(2013),http://www.ideo.com.
[28] Moggridge B. Designing Interactions. Cambridge:MIT Press,2006.
[29] Cagan J,Vogel C M.创造突破性产品——从产品策略到项目定案的创新. 英文版.北京:机械工业出版社,2006.
[30] 何人可.工业设计专业英语. 北京:北京理工大学出版社,1999.
[31] 张伟社,张涛.产品系统设计.西安:陕西科学技术出版社,2006.
[32] 卡尔·T. 尤里齐,斯蒂芬·D,等.产品设计与开发. 杨德林,译. 大连:东北财经大学出版社,2001.
[33] 上海指南工业设计有限公司.产品设计指南. 北京:清华大学出版社,2006.
[34] 尚淼.产品设计形态.武汉:武汉大学出版社,2010.
[35] 李维立.产品设计表达.天津:天津大学出版社,2009.
[36] 王俊涛,肖慧.新产品设计开发. 北京:中国水利水电出版社,2011.